WORKED EXAMPLES
IN MATHEMATICS
for
Scientists & Engineers

WORKED EXAMPLES IN MATHEMATICS

for
Scientists & Engineers

G. STEPHENSON

DOVER PUBLICATIONS, INC.
Mineola, New York

Bibliographical Note

This Dover edition, first published in 2019, is an unabridged and corrected republication of the work originally printed by Longman Group Limited, Essex, England, in 1992 [first publication: 1985].

Library of Congress Cataloging-in-Publication Data

Names: Stephenson, G. (Geoffrey), 1927– author.
Title: Worked examples in mathematics for scientists and engineers / G. Stephenson.
Description: Dover edition. | Mineola, New York : Dover Publications, Inc., 2019. | Originally published: London : Longman, 1985. Republished: Essex, England : Longman, 1992. | Includes index.
Identifiers: LCCN 2019011725| ISBN 9780486837369 | ISBN 048683736X
Subjects: LCSH: Mathematics—Problems, exercises, etc. | Science—Mathematics—Problems, exercises, etc. | Engineering mathematics—Problems, exercises, etc.
Classification: LCC QA43 .S79 2019 | DDC 510.76—dc23
LC record available at https://lccn.loc.gov/2019011725

Manufactured in the United States by LSC Communications
83736X01   2019
www.doverpublications.com

CONTENTS

It is widely accepted that worked examples have an important part to play in the teaching of mathematics, and that they lead to a greater understanding of essential mathematical ideas. Lecture courses usually tend to concentrate, if for no other reason than shortage of time, on the theory rather than examples. This collection of fully worked problems covers most of the topics met in ancillary mathematics courses, and used with a set of lecture notes or conventional textbook should greatly facilitate the understanding of mathematical techniques. It should also provide an effective means of revision for examinations.

The book is especially written for students in scientific disciplines who require mathematical skills, particularly engineers, physicists and chemists. Mathematics students studying mathematical methods as part of their Honours course should, however, find many of these problems of interest, and postgraduate scientists wishing to recall their knowledge of particular topics could well find this book a useful aid.

Most of the examples in this book have been taken from problem sheets I have set for students studying various disciplines at Imperial College. The origins of many of these problems are impossible to identify, and I am therefore grateful to any of my colleagues, past and present, who may have contributed at some time or other to their formulation. Other problems have been chosen from examination papers set for ancillary courses at Imperial College (London University) and have therefore been referred to by (L.U.) in the text. I am grateful to the University for permission to use these problems.; the solutions are, of course, my own responsibility.

I am grateful to Dr. Noel Baker and Norman Froment for reading the manuscript and making a number of valuable comments, and also to my many students who have attempted these problems and shown me more clearly where their difficulties occurred.

Finally it is a pleasure to thank the staff at Longman for their continued friendly cooperation and encouragement.

Imperial College, 1984.

1. FUNCTIONS

[Mappings, even and odd functions, periodic functions, discontinuities, mod x, unit step function, $[x]$]

1.1 Given $f(x) = 2x^2 + x + 3$, calculate the values at $x = 1$ of $f(x - 1)$, $f(x^2)$, $[f(x)]^2$, $f(f(x))$. Show also that $f(x) > 0$.

$f(x - 1) = 2(x - 1)^2 + (x - 1) + 3$. At $x = 1$, $f(x - 1) = f(0) = 3$.
$f(1^2) = f(1) = 6$.
$[f(1)]^2 = 36$.
$f(f(x)) = 2(2x^2 + x + 3)^2 + (2x^2 + x + 3) + 3$. At $x = 1$, therefore, $f(f(1)) = 2(6^2) + 6 + 3 = 81$. $f(x) = 2x^2 + x + 3 = 2(x + \frac{1}{4})^2 + \frac{23}{8}$ which, for real x, is always positive.

1.2 Given $f(x) = \dfrac{x + 1}{x - 1}$ and $g(y) = \dfrac{5y + 2}{2y + 1}$, find $g(f(x))$ and $f(g(x))$.

$g(f(x)) = \dfrac{5[(x + 1)/(x - 1)] + 2}{2[(x + 1)/(x - 1)] + 1} = \dfrac{7x + 3}{3x + 1}$.

Likewise $f(g(x)) = \dfrac{[(5x + 2)/(2x + 1)] + 1}{[(5x + 2)/(2x + 1)] - 1} = \dfrac{7x + 3}{3x + 1}$.

Hence $g(f(x)) = f(g(x))$. (N.B. This is not necessarily so for other forms of f and g. For example, if $f(x) = e^x$, $g(y) = \sin y$, then $f(g(x)) = e^{\sin x}$, whereas $g(f(x)) = \sin(e^x)$.)

1.3 Show that the function inverse to $(x + 1)/(x - 1)$ is the same function.

If $f(x) = (x + 1)/(x - 1) = y$, say, then $x = f^{-1}(y) = (y + 1)/(y - 1)$. Hence $f^{-1}(x) = (x + 1)/(x - 1)$, which is the function $f(x)$.

1.4 Determine the range of the function $f(x) = x^2 + 1$ corresponding to each of the x-domains, and state whether the mapping is 1–1 or many–1: (a) $-1 \leqslant x \leqslant 1$, (b) $2 \leqslant x \leqslant 4$.

The graph of $f(x)$ is shown in Fig. 1.

1

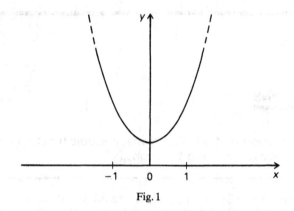

Fig. 1

(a) $f(x)$ lies between 1 (when $x = 0$) and 2 (when $|x| = 1$). Hence range is $1 \leqslant f \leqslant 2$. Mapping is many:1 since if $x = \pm a$, where $a \leqslant 1$, we find the same values of $f(x)$. Mapping is, in fact, 2:1.

(b) $f(2) = 5$, $f(4) = 17$. Function is strictly monotonic increasing in $2 \leqslant x \leqslant 4$. Range is $5 \leqslant f \leqslant 17$. For each x in this domain there is only one value of f. Hence mapping is 1:1.

1.5 Mod x $(= |x|)$ is defined by

$$|x| = \begin{cases} x, & x \geqslant 0 \\ -x, & x < 0 \end{cases}.$$ Draw the graph of $x - |x|$.

For $x \geqslant 0$, $x - |x| = 0$. For $x < 0$, $x - |x| = 2x$ (see Fig. 2).

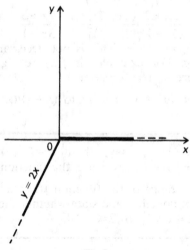

Fig. 2

1.6 Show that the equation $x^2+2y^3=3$ determines y as a single-valued function of x, where x and y are real numbers.

$y = \sqrt[3]{\dfrac{3-x^2}{2}}$. The cube root of a real number (positive or negative) is unique. Hence y is a single-valued function.

1.7 Determine whether the following functions are even, odd or neither:
(a) x^3+6x, (b) $x^2+2\sin x$, (c) $e^{x^2}\cos 3x$.

Even if $f(-x)=f(x)$, odd if $f(-x)=-f(x)$.
(a) $f(x)=x^3+6x$, $f(-x)=-x^3-6x=-f(x)$. Hence odd.
(b) $f(x)=x^2+2\sin x$, $f(-x)=x^2-2\sin x$. Neither even nor odd.
(c) $f(x)=e^{x^2}\cos 3x$, $f(-x)=e^{x^2}\cos 3x = f(x)$. Hence even.

1.8 Write $\sqrt{(x-1)/(x+1)}$ as the sum of an even and an odd function, and state the range of values of x for which the function is defined.

$$\sqrt{\dfrac{x-1}{x+1}}=\sqrt{\dfrac{(x-1)^2}{(x+1)(x-1)}}=\dfrac{x-1}{\sqrt{x^2-1}}=\dfrac{x}{\sqrt{x^2-1}}-\dfrac{1}{\sqrt{x^2-1}}.$$

The first term is an odd function of x, and the second an even function. The function exists for $x^2-1>0$ and $x=1$.

1.9 Find which of the following functions are periodic and state their periods:
(a) $\cos 2x$, (b) $\dfrac{\sin x}{x}$, (c) $|\sin x|$.

$f(x)$ is periodic with period T if $f(x+nT)=f(x)$ for all x ($n=0,1,2,\ldots$).
(a) $\cos 2(x+nT)=\cos 2x$ if $T=\pi$. Hence periodic; period π.
(b) $\sin x/x$ is a sine curve with an amplitude decreasing as $1/x$ (see Fig. 3). Hence not periodic.
(c) $|\sin x|=|\sin(x+nT)|$ if $T=\pi$ (see Fig. 4). Periodic; period π.

1.10 Sketch the graphs of $f(x)=xH(x)$, where $H(x)$ is the unit step function, and of $f(x)=H(x)-H(x-c)+H(x-2c)-\cdots$, where $0\leqslant x<\infty$, and c is a positive constant.

$$H(x-c)=\begin{cases}0, & x<c\\1, & x\geqslant c\end{cases}.$$

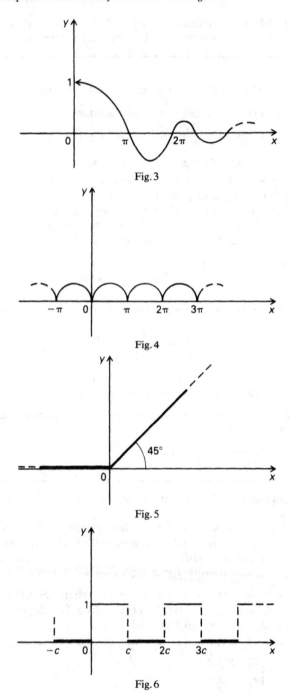

Fig. 3

Fig. 4

Fig. 5

Fig. 6

Hence $f(x) = xH(x) = 0$ for $x < 0$, and equals x for $x \geqslant 0$ (see Fig. 5). The graph of $f(x) = H(x) - H(x-c) + H(x-2c) - \cdots$ is shown in Fig. 6.

1.11 A function $f(x)$ is such that for $x < 0$, $f(x) = x^2$, and for $x \geqslant 0$, $f(x) = x$. Represent $f(x)$ in terms of x, x^2 and $H(x)$.

Consider $x^2 + H(x)(x - x^2)$. Then for $x < 0$ this expression is x^2. For $x \geqslant 0$, expression is x. Hence $f(x) = x^2 + H(x)(x - x^2)$.

1.12 At what points are the following functions discontinuous:

(a) $\dfrac{1}{1-x}$, (b) $\dfrac{x}{x^2-4}$, (c) cosec x.

(a) Function has an infinite discontinuity at $x = 1$.
(b) Function has infinite discontinuities when $x^2 - 4 = 0$, i.e. at $x = \pm 2$.
(c) cosec $x = 1/\sin x$. Now $\sin x = 0$ at $x = n\pi$, where $n = 0, \pm 1, \pm 2, \ldots$. Hence discontinuities at all these values.

1.13 The function $[x]$ is defined as the integer part of x. Sketch the graphs of $[x]$, and of $x - [x]$.

$[0.4] = 0$, $[1.2] = 1$, $[-0.5] = [-1 + 0.5] = -1$, and so on (see Fig. 7). $x - [x] = x$ if $0 < x < 1$, $x - [x] = x - 1$ if $1 < x < 2$, and so on (see Fig. 8). The period of this function is unity.

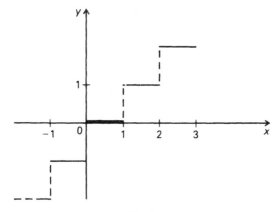

Fig. 7

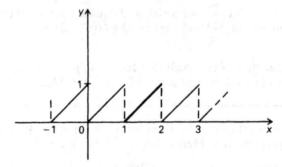

Fig. 8

1.14 If $f(f(x)) = x^2 - 2$, verify that two solutions are

$$f(x) = 2 \cos (\sqrt{2} \cos^{-1} (x/2))$$

and

$$f(x) = 2 \cosh (\sqrt{2} \cosh^{-1} (x/2)).$$

$$\left(\text{N.B. } \cosh x = \frac{e^x + e^{-x}}{2}, \quad \cosh 2x = 2 \cosh^2 x - 1 \right).$$

Consider the first solution. Then

$$f(f(x)) = 2 \cos \left[\sqrt{2} \cos^{-1} \left\{ \tfrac{1}{2} \cdot 2 \cos \left(\sqrt{2} \cos^{-1} \left(\tfrac{x}{2} \right) \right) \right\} \right]$$

$$= 2 \cos \left[\sqrt{2} \cdot \sqrt{2} \cos^{-1} \left(\tfrac{x}{2} \right) \right]$$

$$= 2 \cos \left[2 \cos^{-1} \left(\tfrac{x}{2} \right) \right] = 2 \left[2 \cos^2 \left(\cos^{-1} \left(\tfrac{x}{2} \right) \right) - 1 \right]$$

$$= 4 \left(\tfrac{x}{2} \right)^2 - 2 = x^2 - 2.$$

Similarly for the second solution.

2. INEQUALITIES

[Arithmetic, geometric and harmonic means, induction methods, Stirling's formula]

2.1 Find the ranges of real values of x which satisfy the inequalities

(a) $x^2 + 8 < 2x$, (b) $-1 < \dfrac{3x+4}{x-6} < 1$.

(a) We require $x^2 - 2x + 8 < 0$. But $x^2 - 2x + 8 = (x-1)^2 + 7$, which is always positive. Hence no real x satisfies this inequality.
(b) For $(3x+4)/(x-6) > -1$ we have $(4x-2)/(x-6) > 0$, which is satisfied if $x < \frac{1}{2}$ or $x > 6$.
For $(3x+4)/(x-6) < 1$ we have $(2x+10)/(x-6) < 0$, which requires $6 > x > -5$. Hence for both parts of the inequality to be satisfied we must have $-5 < x < \frac{1}{2}$.

2.2 Show that for $0 < x < 1$, $1 + x < e^x < 1/(1-x)$.

For $0 \leqslant x < 1$, $\dfrac{1}{1-x} = (1-x)^{-1} = 1 + x + x^2 + x^3 + \cdots$, using the binomial series.
Since $e^x = 1 + x + x^2/2! + x^3/3! + \cdots$, it is required to prove that

$$1 + x < 1 + x + x^2/2! + x^3/3! + \cdots < 1 + x + x^2 + x^3 + \cdots.$$

Direct comparison of corresponding terms provides the proof.

2.3 Given that x_1, x_2, x_3 are positive numbers, show that

$$(x_1 + x_2 + x_3)\left(\frac{1}{x_1} + \frac{1}{x_2} + \frac{1}{x_3}\right) \geqslant 9.$$

Arithmetic mean A of $x_1, x_2, x_3, \ldots, x_n$ is $(x_1 + x_2 + \cdots + x_n)/n$, geometric mean $G = \sqrt{x_1 x_2 \cdots x_n}$, and harmonic mean H defined by

$$\frac{1}{H} = \frac{1}{n}\left(\frac{1}{x_1} + \frac{1}{x_2} + \cdots + \frac{1}{x_n}\right).$$

7

Basic result

$A \geqslant G \geqslant H$, equality signs holding only when $x_1 = x_2 = x_3 = \cdots = x_n$. Now using $A \geqslant H$ gives

$$\tfrac{1}{3}(x_1 + x_2 + x_3) \geqslant 3 \Big/ \left(\frac{1}{x_1} + \frac{1}{x_2} + \frac{1}{x_3}\right).$$

Since x_1, x_2, x_3 are positive, each side can be multiplied by the *positive* number $\left(\dfrac{1}{x_1} + \dfrac{1}{x_2} + \dfrac{1}{x_3}\right)$ to give the result.

Alternatively, the left-hand side of the inequality could be written as

$$(x_1 + x_2 + x_3)\left(\frac{1}{x_1} + \frac{1}{x_2} + \frac{1}{x_3}\right)$$

$$= \left(\frac{x_1}{x_2} + \frac{x_2}{x_1}\right) + \left(\frac{x_2}{x_3} + \frac{x_3}{x_2}\right) + \left(\frac{x_3}{x_1} + \frac{x_1}{x_3}\right) + 3.$$

Using $A \geqslant G$,

$$\frac{1}{2}\left(\frac{x_1}{x_2} + \frac{x_2}{x_1}\right) \geqslant \sqrt{\frac{x_1}{x_2} \cdot \frac{x_2}{x_1}} = 1,$$

and likewise for other terms. Hence left-hand side $\geqslant 2 + 2 + 2 + 3 = 9$.

2.4   Find the value of x for which the function $y = \dfrac{a + bx^4}{x^2}$ has its least value, where a and b are positive constants.

Using $A \geqslant G$,

$$\frac{1}{2}\left(\frac{a}{x^2} + bx^2\right) \geqslant \sqrt{\frac{a}{x^2} \cdot bx^2} \qquad \text{or} \qquad y = \frac{a}{x^2} + bx^2 \geqslant 2\sqrt{ab}.$$

The least value occurs when the equality sign holds, which requires that

$$\frac{a}{x^2} = bx^2 \qquad \text{or} \qquad x = \sqrt[4]{\frac{a}{b}}.$$

2.5   Prove by induction that

(a) $\displaystyle\sum_{r=1}^{n} \frac{1}{\sqrt{r}} > \sqrt{n}, \ (n \geqslant 2)$,   (b) $n^n e^{-n} < n! \ (n \geqslant 1)$.

(a) Let $P(n) = \sum_{r=1}^{n} 1/\sqrt{r}$. Then $P(2) = \sum_{r=1}^{2} 1/\sqrt{r} = 1 + 1/\sqrt{2} > \sqrt{2}$. Hence $P(2) > \sqrt{2}$ is true. Now assume that $P(k) = \sum_{r=1}^{k} 1/\sqrt{r} > \sqrt{k}$ is true for some k. Then

$$P(k+1) = \sum_{r=1}^{k+1} \frac{1}{\sqrt{r}} = \sum_{r=1}^{k} \left(\frac{1}{\sqrt{r}}\right) + \frac{1}{\sqrt{k+1}} > \sqrt{k} + \frac{1}{\sqrt{k+1}}$$

$$= \frac{\sqrt{k}\sqrt{k+1}+1}{\sqrt{k+1}} > \frac{k+1}{\sqrt{k+1}} = \sqrt{k+1},$$

so if $P(k) > \sqrt{k}$ is true so also is $P(k+1) > \sqrt{k+1}$. Since $P(2) > \sqrt{2}$ is true, the formula is true for all n.

(b) Let $P(N) = n^n e^{-n}$. Then $P(1) = e^{-1} < 1!$ is true. Assume now that

$$P(k) = k^k e^{-k} < k!$$

is true for some k. Then

$$P(k+1) = \frac{(k+1)^{k+1}}{e^{k+1}} = \frac{k^k}{e^k} \cdot \frac{1}{k^k} \frac{(k+1)^{k+1}}{e} = \frac{k^k}{e^k}(k+1)\frac{(1+1/k)^k}{e}.$$

But the exponential number e is defined by

$$\lim_{k\to\infty}\left(1+\frac{1}{k}\right)^k.$$

Hence $(1+1/k)^k < e$. Finally therefore

$$\frac{k^k}{e^k} \cdot (k+1)\frac{(1+1/k)^k}{e} < \frac{k^k(k+1)}{e^k} < (k+1)k!$$

using $P(k) = ke^{-k} < k!$. Hence $P(k+1) < (k+1)!$. Since $P(1) < 1!$ is true, the formula is true for all n.

2.6 Stirling's approximate formula for $n!$ where n is large is

$$n! \approx \sqrt{2\pi}\, n^{n+1/2} e^{-n}\left(1+\frac{1}{12n}\right).$$

Evaluate $n!$ for $n = 6$ and compare with the exact value.

By Stirling's formula $6! \approx \sqrt{2\pi} \cdot 6^{6.5} e^{-6}(1+\frac{1}{72}) = 719.94$.
The exact value of $6!$ is 720.
(N.B. Stirling's formula, although really only valid for large n, is in fact quite a good approximation even for $n = 3$ since it gives $3! \approx 5.9983$ as compared with the exact value of 6).

3. LIMITS

[Sequences, l'Hôpital's rule, series expansions]

3.1 Determine the limits as $n \to \infty$ (n integral) of the following sequences:

(a) $\dfrac{n}{n+1}$,   (b) $\dfrac{2n}{n^2+1}$.

For the second sequence, find the first term which is within 0.001 of the limit.

(a) $n/(n+1) = 1/(1+1/n)$. As $n \to \infty$, expression tends to unity.
(b) $2n/(n^2+1)$, for large n, behaves like $2/n$. Hence as $n \to \infty$, the limit is zero. If $2n/(n^2+1) \leqslant 0.001$, then $n^2 - 2000n + 1 > 0$, whence $n = 2000$ is the first term which is within 0.001 of the zero limit.

3.2 Evaluate the following limits:

(a) $\displaystyle \lim_{x \to \infty} \left(\frac{x^3+3}{2x^3+4x+1} \right)$,   (b) $\displaystyle \lim_{x \to 0} \left(\frac{1 - \cos x}{x^2} \right)$,

(c) $\displaystyle \lim_{x \to 0} \left(\frac{\tan px}{\tan qx} \right)$,   (d) $\displaystyle \lim_{x \to 1} \left(\frac{1 + \cos \pi x}{\tan^2 \pi x} \right)$.

(a) $\dfrac{x^3+3}{2x^3+4x+1} = \dfrac{1+3/x^3}{2+4/x^2+1/x^3}$. For large x, expression tends to $\frac{1}{2}$.

(b) Maclaurin series expansion of $\cos x$ is $1 - \dfrac{x^2}{2!} + \dfrac{x^4}{4!} - \dfrac{x^6}{6!} + \cdots$ (see 7.1(b)). Hence $(1 - \cos x)/x^2 = \frac{1}{2} - x^2/4 + \cdots$. As $x \to 0$, limit $= \frac{1}{2}$. Alternatively, l'Hôpital's rule states that

$$\lim_{x \to a} \frac{f(x)}{g(x)} = \lim_{x \to a} \frac{f'(x)}{g'(x)}$$

if $f(a) = g(a) = 0$ and the right-hand limit exists (see also 7.7).

10

Putting $f(x) = 1 - \cos x$, and $g(x) = x^2$,

$$\lim_{x \to 0} \left(\frac{1 - \cos x}{x^2} \right) = \lim_{x \to 0} \left[\frac{d(1 - \cos x)/dx}{d(x^2)/dx} \right]$$

$$= \lim_{x \to 0} \left(\frac{\sin x}{2x} \right) = \frac{1}{2} \lim_{x \to 0} \left(\frac{\sin x}{x} \right) = \frac{1}{2},$$

as before (using the basic result $\lim_{x \to 0} (\sin x/x) = 1$, which can also be proved by l'Hôpital's rule).

(c) $\lim_{x \to 0} \left(\frac{\tan px}{\tan qx} \right) = \frac{p}{q} \lim_{x \to 0} \left(\frac{\tan px}{px} \right) \left(\frac{qx}{\tan qx} \right)$

$$= \frac{p}{q} \lim_{x \to 0} \left(\frac{\tan px}{px} \right) \lim_{x \to 0} \left(\frac{qx}{\tan qx} \right)$$

(using the basic property of limits that the limit of a product is the product of the limits). But

$$\lim_{x \to 0} \left(\frac{\tan px}{px} \right) = \lim_{x \to 0} \left(\frac{\sin px}{px} \right) \lim_{x \to 0} \left(\frac{1}{\cos px} \right) = 1.$$

Hence the required limit is p/q.

(d) $\dfrac{1 + \cos \pi x}{\tan^2 \pi x} = \dfrac{\cos^2 \pi x (1 + \cos \pi x)}{\sin^2 \pi x}$

$$= \frac{\cos^2 \pi x (1 + \cos \pi x)}{1 - \cos^2 \pi x} = \frac{\cos^2 \pi x}{1 - \cos \pi x}.$$

Hence

$$\lim_{x \to 1} \left(\frac{1 + \cos \pi x}{\tan^2 \pi x} \right) = \frac{1}{2}.$$

Alternatively, since $1 + \cos \pi x$ and $\tan^2 \pi x$ are both zero at $x = 1$, we may use l'Hôpital's rule to get

$$\lim_{x \to 1} \left(\frac{- \pi \sin \pi x}{2 \pi \tan \pi x \sec^2 \pi x} \right) = \frac{1}{2}$$

as before.

3.3 Evaluate the following limits:

(a) $e^{-x}(4 \sinh^3 x - \cosh 3x)$ as $x \to \infty$,

(b) $x(\sqrt{x^2 - 1} - (x^3 - 1)^{1/3})$ as $x \to \infty$,

(c) $\cos^2 2x / \left(x - \dfrac{\pi}{4} \right)^2$ as $x \to \dfrac{\pi}{4}$,

(d) $x \sin \{(x^2 + 1)/x\}$ as $x \to 0$.   (L.U.)

(a) Using the definitions $\sinh x = (e^x - e^{-x})/2$, $\cosh x = (e^x + e^{-x})/2$, the given function becomes

$$\tfrac{1}{2}e^{-x}\{(e^x - e^{-x})^3 - e^{-3x} - e^{3x}\} = \frac{e^{-x}}{2}\{-3e^x + 3e^{-x} - e^{-3x} - e^{-3x}\},$$

which tends to $-\tfrac{3}{2}$ as $x \to \infty$.
(b) Writing the expression as $x\{x(1-1/x^2)^{1/2} - x(1-1/x^3)^{1/3}\}$ and expanding, we find

$$x\left\{x - \frac{1}{2x} + \cdots - x + \frac{1}{3x^2} + \cdots\right\} \to -\frac{1}{2} \qquad \text{as } x \to \infty.$$

(c) Using l'Hôpital's rule,

$$\lim_{x\to\pi/4}\left[\frac{\cos^2 2x}{(x-\pi/4)^2}\right] = \lim_{x\to\pi/4}\left[\frac{-4\cos 2x \sin 2x}{2(x-\pi/4)}\right]$$

$$= \lim_{x\to\pi/4}\left(-\frac{\sin 4x}{x-\pi/4}\right) = -4\lim_{x\to\pi/4}(\cos 4x) = 4.$$

Alternatively, evaluate

$$\lim_{x\to\pi/4}\left(\frac{\cos 2x}{x-\pi/4}\right)$$

and square the result.
(d) $|x \sin\{(x^2+1)/x\}| \le x$, since the sine function is bounded by ± 1. Hence as $x \to 0$ the expression tends to zero.

3.4 Draw the graph of $\sin(1/x)$.
Evaluate $\lim_{x\to 0}(x \sin(1/x))$ and $\lim_{x\to 0}(x \cos(1/x))$, and hence show that if $f(x) = x^2 \sin(1/x)$ and $g(x) = x^3 \sin(1/x)$, then $\lim_{x\to 0} f'(x)$ does not exist, but $\lim_{x\to 0} g'(x) = 0$.

The graph of $\sin(1/x)$ is shown in Fig. 9. Since $|\sin(1/x)| \le 1$ $(x \ne 0)$,

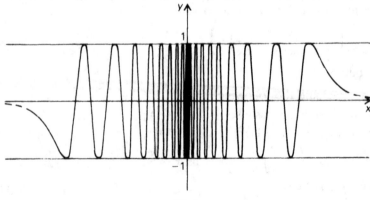

Fig. 9

it follows that $-|x| \le x \sin(1/x) \le |x|$ for all x. Therefore

$$-\lim_{x\to 0}|x| \le \lim_{x\to 0} x \sin\frac{1}{x} \le \lim_{x\to 0}|x|,$$

whence $\lim_{x\to 0}(x \sin 1/x) = 0$. Similarly $\lim_{x\to 0}(x \cos 1/x) = 0$.

$$f(x) = x^2 \sin\frac{1}{x},$$

$$f'(x) = 2x \sin\frac{1}{x} + \left(x^2 \cos\frac{1}{x}\right)\left(-\frac{1}{x^2}\right) = 2x \sin\frac{1}{x} - \cos\frac{1}{x}.$$

The last term oscillates between ± 1 increasingly rapidly as $x \to 0$. Hence the limit does not exist.

$$g(x) = x^3 \sin\frac{1}{x},$$

$$g'(x) = 3x^2 \sin\frac{1}{x} + \left(x^3 \cos\frac{1}{x}\right)\left(-\frac{1}{x^2}\right) = 3x^2 \sin\frac{1}{x} - x \cos\frac{1}{x}.$$

Therefore $\lim_{x\to 0} g'(x) = 0$, using the previous results.

---

4. DIFFERENTIATION

[Continuity, differentiation formulae, higher derivatives, Leibnitz's theorem, maxima and minima]

4.1 Sketch the graph of the function $f(x) = \frac{1}{2}(x + |x|)$ and deduce that, for $x \ne 0$, $f'(x) = H(x)$, where $H(x)$ is the unit step function.

When $x > 0$, $f(x) = \frac{1}{2}(x + x) = x$. When $x < 0$, $f(x) = 0$. The graph is shown in Fig. 10. Clearly $f'(x) = 0$ for $x < 0$, $f'(x) = 1$ for $x > 0$. Hence

$$f'(x) = \begin{Bmatrix} 0, & x < 0 \\ 1, & x > 0 \end{Bmatrix} = H(x)$$

(using the definition given in 1.10).

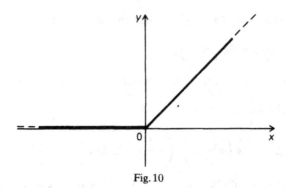

Fig. 10

4.2 Given $f'(x) = \frac{1}{a} f(x-a)$, and if $f(x) = \left(\frac{x}{a}\right)^{1/2}$ for $0 \le x \le a$, prove that

$$f(x) = 1 + \frac{2}{3}\left(\frac{x}{a} - 1\right)^{3/2} \qquad \text{for } a \le x \le 2a$$

assuming continuity at $x = a$.

Since $f(x) = (x/a)^{1/2}$ for $0 \le x \le a$, then $f(x-a) = [(x-a)/a]^{1/2}$ for $0 \le x - a \le a$ or $a \le x \le 2a$. Hence $f'(x) = 1/a[(x-a)/a]^{1/2}$ for $a \le x \le 2a$. Integrating gives $f(x) = A + \frac{2}{3}((x/a)-1)^{3/2}$, where A is an arbitrary constant of integration. At $x = a$, $f(a) = 1$. Hence, assuming continuity, $A = 1$.

4.3 If

$$y = \begin{cases} \dfrac{1}{x^2} \sinh x, & 0 < x \le 2, \\[2mm] Ax + B, & x \ge 2, \end{cases}$$

determine the constants A and B so that both y and dy/dx are continuous.

Continuity of y at $x = 2$ gives $2A + B = \frac{1}{4} \sinh 2$. Continuity of dy/dx at $x = 2$ gives $A = [(\cosh 2)/2^2 - (2 \sinh 2)/2^3]$. Hence $A = e^{-2}/4$ and $B = \frac{1}{4} \sinh 2 - e^{-2}/2$.

4.4 Differentiate with respect to x:
(a) $x^3 \cos x$,   (b) $(2x-1)^{-3/2}$,   (c) $\sqrt{x^2 - 5x + 7}$,

(d) $\log_e (\sec x + \tan x)$,   (e) $\log_e \cos \left(\dfrac{1}{x}\right)$,   (f) $x^{\sin x}$,

(g) $\log_e (\log_e \sin x)$.

Basic results

If u and v are differentiable functions of x, then

$$\frac{d}{dx}(uv) = u\frac{dv}{dx} + v\frac{du}{dx} \qquad \text{(product formula)},$$

$$\frac{d}{dx}\left(\frac{u}{v}\right) = \frac{v\,du/dx - u\,dv/dx}{v^2} \qquad \text{(quotient formula)}.$$

If $y = f(u)$ and $u = u(x)$, then

$$\frac{dy}{dx} = \frac{df}{du} \cdot \frac{du}{dx} \qquad \text{(function of a function formula)}.$$

(a) By the product formula, the differential coefficient is $3x^2 \cos x - x^3 \sin x$.

(b) Let $2x - 1 = u$. The differential coefficient is $-3(2x-1)^{-5/2}$.

(c) Let $x^2 - 5x + 7 = u$. The differential coefficient is $\dfrac{1}{2}\dfrac{(2x-5)}{\sqrt{x^2-5x+7}}$.

(d) Let $\sec x + \tan x = u$. Then

$$\frac{d}{dx}\log_e(\sec x + \tan x) = \frac{d}{du}(\log_e u)\frac{du}{dx}$$

$$= \frac{1}{u}(\sec x \tan x + \sec^2 x) = \sec x.$$

(e) Let $\cos(1/x) = u$. Then

$$\frac{d}{dx}\log_e \cos\left(\frac{1}{x}\right) = \frac{d}{du}(\log_e u)\frac{du}{dx}$$

$$= \frac{1}{u}\left(-\sin\left(\frac{1}{x}\right)\right)\left(-\frac{1}{x^2}\right) = \frac{1}{x^2}\tan\left(\frac{1}{x}\right).$$

(f) Let $f(x) = x^{\sin x}$. Take the logarithm of each side: $\log_e f(x) = \sin x \log_e x$. Hence

$$\frac{1}{f}\frac{df}{dx} = \frac{\sin x}{x} + \cos x \log_e x.$$

Therefore

$$\frac{df}{dx} = x^{\sin x}\left(\frac{\sin x}{x} + \cos x \log_e x\right).$$

(g) Function not defined for real x since $|\sin x| \leqslant 1$, $\log_e \sin x \leqslant 0$, the logarithm of a negative number being undefined in the real number system.

15

4.5 Differentiate (a) a^x, (b) $\log_a x$.

(a) $y = a^x$, $\log_e y = x \log_e a$. Hence $\dfrac{1}{y}\dfrac{dy}{dx} = \log_e a$, $\dfrac{dy}{dx} = a^x \log_e a$.

(b) $y = \log_a x$, $a^y = x$. Hence $\dfrac{dx}{dy} = a^y \log_e a$, $\dfrac{dy}{dx} = \dfrac{1}{x \log_e a}$.

4.6 Find dy/dx in terms of x and y when $x = at/(1-t^3)$, $y = at^2/(1-t^3)$, where a is a constant.

$$\frac{dy}{dx} = \frac{dy/dt}{dx/dt} = \left[\frac{(1-t^3)2at+3at^4}{(1-t^3)^2}\right]\bigg/\left[\frac{(1-t^3)a+3at^3}{(1-t^3)^2}\right] = \frac{t(2+t^3)}{1+2t^3}.$$

From the definitions of x and y, $t = y/x$. Hence

$$\frac{dy}{dx} = \left(\frac{y}{x}\right)\left(\frac{2x^3+y^3}{x^3+2y^3}\right).$$

4.7 Express d^2x/dy^2 in terms of derivatives of y with respect to x.

$$\frac{dx}{dy} = \frac{1}{(dy/dx)}.$$

$$\frac{d^2x}{dy^2} = \frac{d}{dy}\left(\frac{dx}{dy}\right) = \frac{d}{dx}\left(\frac{1}{dy/dx}\right)\frac{dx}{dy} = -\frac{1}{(dy/dx)^2}\frac{d^2y}{dx^2}\frac{dx}{dy}$$

$$= -\frac{1}{(dy/dx)^3}\frac{d^2y}{dx^2}.$$

4.8 Given that D^n denotes the operator $\dfrac{d^n}{dx^n}$, show that

(a) $D^n(\sin ax) = a^n \sin(ax + \tfrac{1}{2}n\pi)$ (where a is a constant),
(b) $D^n(x^3 \log_e x) = 6(-1)^n(n-4)!\, x^{-n+3}$ $(n \geqslant 4)$.

(a) If $n = 1$, $D(\sin ax) = a \cos ax = a \sin(ax + \pi/2)$. Hence formula is true. Consider $D^k(\sin ax) = a^k \sin(ak + \tfrac{1}{2}k\pi)$ to be true for some k. Then

$$D^{k+1}(\sin ax) = a^{k+1} \cos(ax + \tfrac{1}{2}k\pi) = a^{k+1} \sin(ax + \tfrac{1}{2}(k+1)\pi),$$

which is true also. Hence since the formula is true for $n = 1$, it is true for all n.

Basic result

Leibnitz's formula states that

$$D^n(uv) = uD^nv + (^nC_1)Du \cdot D^{n-1}v + (^nC_2)D^2u \cdot D^{n-2}v + \cdots + vD^nu,$$

where u and v are differentiable functions of x [N.B. $^nC_r = n!/[r!\,(n-r)!]$]. Now from first principles,

$$D^n(\log_e x) = \frac{(-1)^{n+1}(n-1)!}{x^n} \qquad \text{for } n \geqslant 1.$$

Hence using Leibnitz's theorem

$$D^n(x^3 \log_e x) = x^3 \frac{(-1)^{n+1}(n-1)!}{x^n} + (^nC_1)\cdot\frac{3x^2(-1)^n(n-2)!}{x^{n-1}}$$

$$+ (^nC_2)\cdot\frac{6x(-1)^{n-1}(n-3)!}{x^{n-2}} + (^nC_3)\cdot\frac{6\cdot(-1)^{n-2}(n-4)!}{x^{n-3}}$$

$$= \frac{6(-1)^n(n-4)!}{x^{n-3}} \qquad (n \geqslant 4).$$

4.9 Prove that $D^n\!\left(\dfrac{e^x}{x}\right) = (-1)^n n!\,\dfrac{e^x}{x^{n+1}}\,P_n(-x)$, where $P_n(x)$ is the polynomial obtained by taking the first $n+1$ terms in the Maclaurin expansion of e^x.

It is easily seen that $D^n(1/x) = (-1)^n(n!/x^{n+1})$. Using Leibnitz's theorem

$$D^n\!\left(e^x\frac{1}{x}\right) = e^x(-1)^n\frac{n!}{x^{n+1}} + e^x n(-1)^{n-1}\frac{(n-1)!}{x^n},$$

$$+ e^x\frac{n(n-1)}{2!}\frac{(-1)^{n-2}(n-2)!}{x^{n-1}} + \cdots + \frac{e^x}{x}$$

$$= e^x(-1)^n\frac{n!}{x^{n+1}}\left(1 - x + \frac{x^2}{2!} - \cdots + (-1)^n\frac{x^n}{n!}\right)$$

$$= e^x(-1)^n\frac{n!}{x^{n+1}}\,P_n(-x).$$

4.10 Find the stationary points of the function $y = (2x^2-5x-25)/(x^2+x-2)$ and determine their nature. Locate the points of discontinuity, and evaluate $\lim_{x\to\pm\infty} y$.

$$\frac{dy}{dx} = \frac{7(x+1)(x+5)}{(x^2+x-2)^2}.$$

Stationary points where $dy/dx = 0$. Hence $x = -1$ and $x = -5$ are such points. By evaluating d^2y/dx^2 we find that $x = -5$ gives a maximum, and $x = -1$ a minimum. The function is discontinuous

17

where $x^2+x-2=0$, that is, when $x=1$ and -2.

$$\lim_{x\to\pm\infty}\left\{\frac{2x^2-5x-25}{x^2+x-2}\right\}=\lim_{x\to\pm\infty}\left\{\frac{2-\dfrac{5}{x}-\dfrac{25}{x^2}}{1+\dfrac{1}{x}-\dfrac{2}{x^2}}\right\}=2.$$

4.11 Find the stationary points of the function $y=x^2e^{-x}$ and determine whether they are maxima or minima.

Now $dy/dx=e^{-x}x(2-x)$. Hence the stationary points are at $x=0$ and $x=2$. $d^2y/dx^2=-xe^{-x}$, so $y''<0$ at $x=2$. Hence a maximum occurs at $x=2$. At $x=0$, $y''=0$, so we need to examine the next *even* order differential coefficient at $x=0$. Accordingly $d^4y/dx^4=(2-x)e^{-x}>0$ at $x=0$, and therefore a minimum occurs at this point.

4.12 Determine the maxima and minima (if any) of
(a) $\sin^{-1}(x^2+2)$, (b) $1+x^{2/3}$, (c) $\sqrt{1-x^2}+\sqrt{x^2-1}$.

(a) For real x, $x^2+2>1$. Hence $\sin^{-1}(x^2+2)$ does not exist.
(b) Differentiating gives $\frac{2}{3}x^{-1/3}$, which tends to zero as $x\to\pm\infty$. Hence no stationary points. The function possesses a minimum (a *least* value but not a stationary point) at $x=0$. (See Fig. 11.)
(c) The function exists only for $|x|=1$.

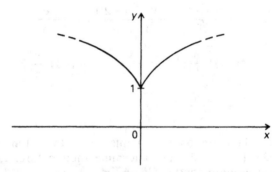

Fig. 11

5. INTEGRATION

[Substitutions, partial fractions, integration by parts, recurrence relations, improper integrals, Cauchy principal value, error function, Γ-function, differentiation of integrals]

5.1 Determine the integrals of

(a) $\dfrac{2x+3}{x^2+3x-2}$, (b) $\sqrt{(1-4x^2)}$, (c) $\sec x$, (d) $\dfrac{\cos x}{3+\cos x}$.

(a) $\displaystyle\int \frac{2x+3}{x^2+3x-2}\,dx = \int \frac{d(x^2+3x-2)}{x^2+3x-2} = \int \frac{du}{u}$, where $u = x^2+3x-2$.

Hence integral is $\log_e (x^2+3x-2)+C$, where C is an integration constant.

(b) $\int \sqrt{(1-4x^2)}\,dx$. Put $x = \tfrac{1}{2}\sin\theta$, $dx = \tfrac{1}{2}\cos\theta\,d\theta$.

$$\int \cos\theta\,(\tfrac{1}{2}\cos\theta)\,d\theta = \frac{1}{2}\int \cos^2\theta\,d\theta = \frac{1}{2}\int \frac{(1+\cos 2\theta)}{2}\,d\theta$$

$$= \tfrac{1}{4}\theta + \tfrac{1}{8}\sin 2\theta + C$$

$$= \tfrac{1}{4}\sin^{-1} 2x + \frac{x}{2}\sqrt{(1-4x^2)} + C.$$

(c) $\displaystyle\int \sec x\,dx = \int \frac{dx}{\cos x} = \int \frac{\cos x}{\cos^2 x}\,dx = \int \frac{d(\sin x)}{1-\sin^2 x}.$ Let $\sin x = t.$

Then

$$\int \frac{dt}{1-t^2} = \frac{1}{2}\int \frac{dt}{1-t} + \frac{1}{2}\int \frac{dt}{1+t}$$

using partial fractions. Hence integral is

$$-\tfrac{1}{2}\log_e (1-t) + \tfrac{1}{2}\log_e (1+t) = \tfrac{1}{2}\log_e \left(\frac{1+t}{1-t}\right) = \tfrac{1}{2}\log_e \left(\frac{1+\sin x}{1-\sin x}\right)$$

$$\int \sec x\,dx = \log_e (\sec x + \tan x) + C.$$

(d) Let $t = \tan(x/2)$, then $dx = 2\,dt/(1+t^2)$, $\cos x = (1-t^2)/(1+t^2)$.

19

Integral is

$$\int \frac{\frac{1-t^2}{1+t^2} \cdot \frac{2dt}{1+t^2}}{3+(1-t^2)/(1+t^2)} = \int \frac{(1-t^2)\,2dt}{(4+2t^2)(1+t^2)} = \int \left(-\frac{3}{2+t^2} + \frac{2}{1+t^2} \right) dt$$

by partial fractions. Hence integral is

$$2 \tan^{-1} t - \frac{3}{\sqrt{2}} \tan^{-1} \left(\frac{t}{\sqrt{2}} \right) + C$$

$$= 2 \tan^{-1} \left(\tan \frac{x}{2} \right) - \frac{3}{\sqrt{2}} \tan^{-1} \left(\frac{1}{\sqrt{2}} \tan \frac{x}{2} \right) + C.$$

5.2 Evaluate by integration by parts

(a) $e^{-x} \cos x$, (b) $x^k \log_e x$, $k \neq 1$, (c) $\dfrac{\log_e x}{(1-x)^2}$.

Basic result

If u and v are functions of x, $\displaystyle\int u \frac{dv}{dx} dx = uv - \int v \frac{du}{dx}$.

(a) $\displaystyle\int e^{-x} \cos x \, dx = e^{-x} \sin x + \int e^{-x} \sin x \, dx$

$$= e^{-x} \sin x + \left[e^{-x}(-\cos x) - \int e^{-x} \cos x \, dx \right].$$

Hence $2 \int e^{-x} \cos x \, dx = e^{-x}(\sin x - \cos x)$, and therefore

$$\int e^{-x} \cos x \, dx = \tfrac{1}{2} e^{-x}(\sin x - \cos x) + C.$$

(b) $\displaystyle\int x^k \log_e x \, dx = \frac{x^{k+1}}{k+1} \log_e x - \int \frac{x^{k+1}}{k+1} \cdot \frac{1}{x} dx$

$$= \frac{x^{k+1}}{k+1} \log_e x - \frac{x^{k+1}}{(k+1)^2} + C.$$

(c) $\displaystyle\int \frac{\log_e x}{(1-x)^2} dx = (\log_e x)\left(\frac{1}{1-x} \right) - \int \frac{1}{x(1-x)} dx$

$$= (\log_e x)\left(\frac{1}{1-x} \right) - \int \left(\frac{1}{x} + \frac{1}{1-x} \right) dx$$

$$= (\log_e x)\left(\frac{1}{1-x} \right) + \log_e \left(\frac{1-x}{x} \right) + C.$$

5.3 Evaluate the definite integrals

(a) $\displaystyle\int_1^2 (x-1)^2 \log_e x \, dx,$ (b) $\displaystyle\int_0^{\pi/2} \frac{dx}{(\sin x + \cos x)^2},$

(c) $\displaystyle\int_0^1 \sqrt{\left(\frac{1+x}{1-x}\right)}\, dx,$ (d) $\displaystyle\int_1^2 \frac{dx}{(4-x)\sqrt{(x-1)}}.$

(a) By parts:

$$\int_1^2 (x-1)^2 \log_e x \, dx = \left[\frac{(x-1)^3}{3}\log_e x\right]_1^2 - \frac{1}{3}\int_1^2 (x-1)^3 \frac{1}{x}\, dx$$

$$= \tfrac{1}{3}\log_e 2 - \frac{1}{3}\int_1^2 (x-1)^3 \frac{dx}{x}$$

$$= \tfrac{1}{3}\log_e 2 - \frac{1}{3}\int_1^2 \left(x^2 - 3x + 3 - \frac{1}{x}\right) dx$$

$$= \tfrac{2}{3}\log_e 2 - \tfrac{5}{18}.$$

(b) Let $\tan x = u$.

$$\int_0^{\pi/2} \frac{dx}{(\sin x + \cos x)^2} = \int_0^{\pi/2} \frac{d(\tan x)}{(\tan x + 1)^2} = \int_0^{\infty} \frac{du}{(u+1)^2}$$

$$= -\left(\frac{1}{1+u}\right)_0^{\infty} = 1.$$

(c) Let $x = \sin t$.

$$\int_0^1 \sqrt{\left(\frac{1+x}{1-x}\right)}\, dx = \int_0^{\pi/2} \sqrt{\frac{1+\sin t}{1-\sin t}}\, \cos t \, dt$$

$$= \int_0^{\pi/2} \sqrt{\frac{(1+\sin t)^2}{1-\sin^2 t}}\, \cos t \, dt = \int_0^{\pi/2} \frac{1+\sin t}{\cos t}\, \cos t \, dt$$

$$= (t - \cos t)_0^{\pi/2} = \frac{\pi}{2} + 1.$$

(d) Let $x = 1+u^2$. The integral now becomes

$$\int_0^1 \frac{2u\, du}{(3-u^2)u} = 2\int_0^1 \frac{du}{3-u^2}$$

$$= 2\cdot\frac{1}{2\sqrt 3}\left[\log_e\left(\frac{\sqrt 3 + u}{\sqrt 3 - u}\right)\right]_0^1 = \frac{1}{\sqrt 3}\log_e\left(\frac{\sqrt 3 + 1}{\sqrt 3 - 1}\right).$$

5.4 Given that $I_n = \int_0^{\pi/2} e^{\alpha x} \cos^n x \, dx$, show that, if $n > 1$,

$$I_n = \frac{n(n-1)}{n^2 + \alpha^2} I_{n-2} - \frac{\alpha}{n^2 + \alpha^2} \qquad (\alpha \text{ is constant}).$$

By parts:

$$I_n = \left(\frac{e^{\alpha x}}{\alpha} \cos^n x \right)_0^{\pi/2} + \int_0^{\pi/2} \frac{e^{\alpha x}}{\alpha} n \cos^{n-1} x \sin x \, dx$$

$$= -\frac{1}{\alpha} + \frac{n}{\alpha^2} \left[(e^{\alpha x} \cos^{n-1} x \sin x)_0^{\pi/2} \right.$$

$$\left. - \int_0^{\pi/2} e^{\alpha x} \{ -(n-1) \cos^{n-2} x \sin^2 x + \cos^{n-1} x \cos x \} \, dx \right]$$

$$= -\frac{1}{\alpha} - \frac{n}{\alpha^2} \int_0^{\pi/2} e^{\alpha x} \{ -(n-1) \cos^{n-2} x (1 - \cos^2 x) + \cos^n x \} \, dx$$

$$= -\frac{1}{\alpha} - \frac{n}{\alpha^2} \int_0^{\pi/2} e^{-\alpha x} \{ -(n-1) \cos^{n-2} x + n \cos^n x \} \, dx$$

$$= -\frac{1}{\alpha} - \frac{n}{\alpha^2} [-(n-1) I_{n-2} + n I_n].$$

Hence $I_n = \dfrac{n(n-1)}{n^2 + \alpha^2} I_{n-2} - \dfrac{\alpha}{n^2 + \alpha^2}.$

5.5 Prove that

$$\int_0^\infty x^n e^{-x^2} \, dx = \tfrac{1}{2}(n-1) \int_0^\infty x^{n-2} e^{-x^2} \, dx,$$

where $n > 1$, and hence evaluate $\int_0^\infty x^5 e^{-x^2} \, dx$.

$$I_{n-2} = \int_0^\infty x^{n-2} e^{-x^2} \, dx = \left(\frac{x^{n-1}}{n-1} e^{-x^2} \right)_0^\infty - \frac{1}{(n-1)} \int_0^\infty x^{n-1}(-2x) e^{-x^2} \, dx$$

$$= \frac{2}{(n-1)} \int_0^\infty x^n e^{-x^2} \, dx = \frac{2 I_n}{(n-1)}.$$

Hence result.

$$I_5 = \tfrac{1}{2} \cdot 4 I_3 = \tfrac{1}{2} \cdot 4 \cdot \tfrac{1}{2} \cdot 2 I_1 = 2 I_1 = 2 \int_0^\infty x e^{-x^2} \, dx = \int_0^\infty e^{-u} \, du \quad (u = x^2)$$

$$= 1.$$

5.6 Prove that

$$\int_0^{\pi/4} \log_e \cos y \, dy = \int_{\pi/4}^{\pi/2} \log_e \sin y \, dy,$$

and hence by putting $x = 2y$, evaluate $\int_0^{\pi/2} \log_e \sin x \, dx$.

Put $y = \pi/2 - u$.

$$\int_0^{\pi/4} \log_e \cos y \, dy = \int_{\pi/2}^{\pi/4} \log_e \sin u(-du) = \int_{\pi/4}^{\pi/2} \log_e \sin u \, du.$$

Now

$$\int_0^{\pi/2} \log_e \sin x \, dx = 2 \int_0^{\pi/4} \log_e (\sin 2y) \, dy$$

$$= 2 \int_0^{\pi/4} (\log_e 2 + \log_e \sin y + \log_e \cos y) \, dy$$

$$= 2 \cdot \frac{\pi}{4} \cdot \log_e 2 + 2 \left(\int_0^{\pi/4} \log_e \sin y \, dy \right.$$

$$\left. + \int_{\pi/4}^{\pi/2} \log_e \sin y \, dy \right) \quad \text{(using the first result).}$$

Therefore (writing the dummy variable y as x)

$$\int_0^{\pi/2} \log_e \sin x \, dx = \frac{\pi}{2} \log_e 2 + 2 \int_0^{\pi/2} \log_e \sin x \, dx$$

or

$$\int_0^{\pi/2} \log_e \sin x \, dx = -\frac{\pi}{2} \log_e 2.$$

5.7 Determine which of the following integrals converge, and find their values when convergent:

(a) $\displaystyle\int_0^\infty \cos x \, dx$, (b) $\displaystyle\int_4^\infty \frac{dx}{(x-2)^2}$, (c) $\displaystyle\int_1^\infty \frac{dx}{(x-2)^2}$,

(d) $\displaystyle\int_0^\infty \frac{e^{-x}}{x} \, dx$, (e) $\displaystyle\int_0^1 x \log_e x \, dx$, (f) $\displaystyle\int_0^1 \frac{dx}{x^\lambda}$.

(a) Divergent since $\sin x$ at the upper limit is not defined.
(b) Convergent: integral is $-[1/(x-2)]_4^\infty = \frac{1}{2}$.
(c) Divergent: integral diverges at $x = 2$.
(d) Divergent: series expansion of integrand gives a $1/x$ term which, when integrated, leads to a logarithmic divergence at $x = 0$.
(e) Convergent: integration by parts gives

$$\left(\frac{x^2 \log_e x}{2} \right)_0^1 - \int_0^1 \frac{x^2}{2} \frac{1}{x} \, dx = -\frac{1}{4}$$

(since $x^2 \log_e x \to 0$ as $x \to 0$).
(f) Divergent if $\lambda \geqslant 1$, convergent if $\lambda < 1$.

5.8 Show that the Cauchy principal value of the improper integral

$$\int_{-1}^{2} \frac{dx}{x} = \log_e 2.$$

Integrand does not exist at $x = 0$, which is in the range of integration. If the integral is to exist then both $\lim_{\varepsilon \to 0} \int_{-1}^{-\varepsilon} \frac{dx}{x}$ and $\lim_{\varepsilon \to 0} \int_{\varepsilon}^{2} \frac{dx}{x}$ must exist. But they both diverge. However, Cauchy principal value defined by

$$\lim_{\varepsilon \to 0} \left[\int_{-1}^{-\varepsilon} \frac{dx}{x} + \int_{\varepsilon}^{2} \frac{dx}{x} \right]$$

exists, and has the value

$$\lim_{\varepsilon \to 0} [\log_e \varepsilon + \log_e 2 - \log_e \varepsilon] = \log_e 2.$$

5.9 Show that the Cauchy principal value of the improper integral

$$\int_{0}^{2} \frac{dx}{1-x^2} = \frac{1}{2} \log_e 3.$$

Integrand does not exist at $x = 1$.
The Cauchy principal value is

$$\lim_{\varepsilon \to 0} \left[\int_{0}^{1-\varepsilon} \frac{dx}{1-x^2} + \int_{1+\varepsilon}^{2} \frac{dx}{1-x^2} \right] = \lim_{\varepsilon \to 0} \left[\int_{0}^{1-\varepsilon} \frac{dx}{1-x^2} - \int_{1+\varepsilon}^{2} \frac{dx}{x^2-1} \right]$$

$$= \lim_{\varepsilon \to 0} \left[\frac{1}{2} \left\{ \log_e \left(\frac{1+x}{1-x} \right) \right\}_{0}^{1-\varepsilon} - \frac{1}{2} \left\{ \log_e \left(\frac{x-1}{x+1} \right) \right\}_{1+\varepsilon}^{2} \right].$$

$$= \frac{1}{2} \lim_{\varepsilon \to 0} \left[\log_e \left(\frac{2-\varepsilon}{\varepsilon} \right) - \log_e \tfrac{1}{3} + \log_e \left(\frac{\varepsilon}{2+\varepsilon} \right) \right]$$

$$= \tfrac{1}{2} \log_e 3.$$

5.10 Evaluate $I = \int_{0}^{1} \sqrt{\log_e \left(\frac{1}{x} \right)} \, dx$ using the Γ-function.

Gamma function defined by $\Gamma(x+1) = \int_0^\infty t^x e^{-t} \, dt$. Integration by parts gives

$$\Gamma(x+1) = [-t^x e^{-t}]_0^\infty + x \int_{0}^{\infty} t^{x-1} e^{-t} \, dt = x\Gamma(x).$$

In I put $x = e^{-t}$. Then

$$I = \int_0^\infty t^{1/2} e^{-t}\, dt = \Gamma(\tfrac{3}{2}) = \tfrac{1}{2}\Gamma(\tfrac{1}{2}).$$

$$\Gamma(\tfrac{1}{2}) = 2\int_0^\infty e^{-u^2}\, du = \sqrt{\pi} \quad \text{(standard integral)}.$$

Hence $I = \dfrac{\sqrt{\pi}}{2}$.

5.11 The error function

$$\text{erf } x = \frac{2}{\sqrt{\pi}} \int_0^x e^{-t^2}\, dt = 1 - \frac{2}{\sqrt{\pi}} \int_x^\infty e^{-t^2}\, dt,$$

since $\int_0^\infty e^{-t^2}\, dt = \sqrt{\pi}/2$. Show that erf x may be expressed as

$$1 - \frac{e^{-x^2}}{x\sqrt{\pi}} \left(1 - \frac{1}{2x^2} + \frac{1 \cdot 3}{(2x)^2} - \frac{1 \cdot 3 \cdot 5}{(2x^2)^3} + \ \cdots \ + R_n \right),$$

where R_n is the remainder after n terms.

Let $e^{-t^2} = \dfrac{1}{t}\dfrac{d}{dt}(-\tfrac{1}{2}e^{-t^2})$. Then

$$\int_x^\infty e^{-t^2}\, dt = \int_x^\infty \frac{1}{t}\frac{d}{dt}(-\tfrac{1}{2}e^{-t^2})\, dt$$

$$= \left[\frac{1}{t}(-\tfrac{1}{2}e^{-t^2}) \right]_x^\infty - \int_x^\infty (-\tfrac{1}{2}e^{-t^2})\left(-\frac{1}{t^2}\right) dt \quad \text{(by parts)}$$

$$= \frac{e^{-x^2}}{2x} + \int_x^\infty (-\tfrac{1}{2}e^{-t^2})\frac{1}{t^2}\, dt = \frac{e^{-x^2}}{2x} - \frac{1}{2}\int_x^\infty \frac{1}{t^3}\frac{d}{dt}(-\tfrac{1}{2}e^{-t^2})\, dt$$

$$= \frac{e^{-x^2}}{2x} - \frac{1}{2}\left[\left\{\frac{1}{t^3}(-\tfrac{1}{2}e^{-t^2})\right\}_x^\infty - \int_x^\infty \frac{3}{t^4}e^{-t^2}\, dt \right]$$

$$= \frac{e^{-x^2}}{2x} - \frac{e^{-x^2}}{4x^3} + \frac{3}{4}\int_x^\infty \frac{1}{t^4}e^{-t^2}\, dt, \text{ and so on.}$$

Hence

$$\text{erf } x = 1 - \frac{2}{\sqrt{\pi}} \int_x^\infty e^{-t^2}\, dt = 1 - \frac{e^{-x^2}}{x\sqrt{\pi}}\left(1 - \frac{1}{2x^2} + \frac{1 \cdot 3}{(2x^2)^2} - \cdots + R_n \right),$$

5.12 Show that if $I(a, b) = \int_0^\infty e^{-ax^2} \cos bx \, dx$, $a > 0$, then $\dfrac{\partial I}{\partial b} = -\dfrac{b}{2a} I$. Hence, using the result that $\int_0^\infty e^{-ax^2} \, dx = \dfrac{1}{2}\sqrt{\dfrac{\pi}{a}}$, show that

$$I(a, b) = \frac{1}{2}\sqrt{\frac{\pi}{a}}\, e^{-b^2/(4a)}.$$

$$\frac{\partial I}{\partial b} = \int_0^\infty e^{-ax^2}(-x \sin bx)\, dx$$

$$= \left(\frac{e^{-ax^2}\sin bx}{2a}\right)_0^\infty - \frac{b}{2a}\int_0^\infty e^{-ax^2}\cos bx\, dx = -\frac{b}{2a} I.$$

Integrating with respect to b, $I(a, b) = Ae^{-b^2/(4a)}$. When $b = 0$

$$I = \int_0^\infty e^{-ax^2}\, dx = \frac{1}{2}\sqrt{\frac{\pi}{a}}.$$

Hence

$$I(a, b) = \frac{1}{2}\sqrt{\frac{\pi}{a}}\, e^{-b^2/(4a)}.$$

5.13 By considering

$$I(\alpha) = \int_0^\infty \frac{e^{-\alpha x}\sin x}{x}\, dx \quad (\alpha \geq 0),$$

show that

$$I(0) = \int_0^\infty \frac{\sin x}{x}\, dx = \frac{\pi}{2}.$$

$$\frac{dI}{d\alpha} = -\int_0^\infty e^{-\alpha x}\sin x\, dx = -\frac{1}{1+\alpha^2} \quad \text{(by parts)}.$$

Hence $I = -\int \dfrac{d\alpha}{1+\alpha^2} = C - \tan^{-1}\alpha$. Now when $\alpha \to \infty$, $I(\alpha) \to 0$, so $0 = C - \tan^{-1}\infty$. Hence $C = \dfrac{\pi}{2}$. When $\alpha = 0$, $I(0) = \int_0^\infty \dfrac{\sin x}{x}\, dx = \dfrac{\pi}{2} - \tan^{-1}0 = \dfrac{\pi}{2}$.

5.14 Show that

$$\frac{d}{dx}\int_a^x f(u)\, du = f(x),$$

where $f(x)$ is any continuous function of x and a is a constant. Hence, or otherwise, find the value of x that maximizes the definite integral

$$\int_0^x \frac{1-u}{\sqrt{1+u}}\,du$$

and evaluate the integral for this value of x. (L.U.)

Basic result

If $I(x) = \int_a^b f(u, x)\,du$, where the upper and lower limits of integration, b and a respectively, are functions of a parameter x, then

$$\frac{dI}{dx} = f(b, x)\frac{db}{dx} - f(a, x)\frac{da}{dx} + \int_a^b \frac{\partial}{\partial x} f(u, x)\,du.$$

From this result

$$\frac{d}{dx}\int_a^x f(u)\,du = f(x)\cdot\frac{dx}{dx} - 0 + 0 = f(x).$$

If $f(u) = (1-u)/\sqrt{1+u}$ then $dI/dx = (1-x)/\sqrt{1+x} = 0$ for a stationary value. Hence $x = 1$ is a stationary point.

Now $\dfrac{d^2 I}{dx^2} = -(1+x)^{1/2} - \frac{1}{2}(1-x)(1+x)^{-3/2}$, which is <0 at $x = 1$.

Therefore $x = 1$ gives the maximum value of I.

$$I(1) = \int_0^1 \frac{1-u}{\sqrt{1+u}}\,du.$$

Put $1 + u = t^2$, $du = 2t\,dt$. Then

$$I(1) = 2\int_1^{\sqrt{2}} (2 - t^2)\,dt = \tfrac{2}{3}(4\sqrt{2} - 5).$$

5.15 Show that $\dfrac{d}{dx}\displaystyle\int_0^x e^{-u}(x-u)^{3/2}\,du = \dfrac{3}{2}\displaystyle\int_0^x e^{-u}\sqrt{x-u}\,du$ *for* $x > 0$

Using the basic result given in 5.14

$$\frac{d}{dx}\int_0^x e^{-u}(x-u)^{3/2}\,du = e^{-x}(x-x)^{3/2}\frac{dx}{dx} - 0 + \int_0^x \frac{\partial}{\partial x}\{e^{-u}(x-u)^{3/2}\}\,du$$

$$= \frac{3}{2}\int_0^x e^{-u}\sqrt{x-u}\,du.$$

5.16 Evaluate $\displaystyle\int_0^x \frac{e^{-x^2}}{(x^2+\frac{1}{2})^2}\,dx$.

Now $\displaystyle\frac{1}{(x^2+\frac{1}{2})^2} = -\frac{1}{2x}\frac{d}{dx}\left(\frac{1}{x^2+\frac{1}{2}}\right)$.

Hence

$$\int\frac{e^{-x^2}}{(x^2+\frac{1}{2})^2}\,dx = -\int\frac{e^{-x^2}}{2x}\frac{d}{dx}\left(\frac{1}{x^2+\frac{1}{2}}\right)dx$$

$$= -\left[\frac{e^{-x^2}}{2x}\frac{1}{x^2+\frac{1}{2}} - \int\frac{1}{x^2+\frac{1}{2}}\frac{d}{dx}\left(\frac{e^{-x^2}}{2x}\right)dx\right] \quad \text{(by parts)}$$

$$= -\left[\frac{e^{-x^2}}{2x(x^2+\frac{1}{2})} - \int\frac{1}{(x^2+\frac{1}{2})}\left\{\frac{-4x^2e^{-x^2}-2e^{-x^2}}{4x^2}\right\}dx\right]$$

$$= -\left[\frac{e^{-x^2}}{2x(x^2+\frac{1}{2})} + \int\frac{e^{-x^2}}{(x^2+\frac{1}{2})}\left(1+\frac{1}{2x^2}\right)dx\right]$$

$$= -\left[\frac{e^{-x^2}}{2x(x^2+\frac{1}{2})} + \int\frac{e^{-x^2}}{x^2}\,dx\right]$$

$$= -\left[\frac{e^{-x^2}}{2x(x^2+\frac{1}{2})} + \left\{-\frac{e^{-x^2}}{x} - 2\int e^{-x^2}\,dx\right\}\right].$$

Inserting the limits therefore

$$\int_0^x \frac{e^{-x^2}}{(x^2+\frac{1}{2})^2}\,dx = 2\int_0^x e^{-x^2}\,dx + \frac{xe^{-x^2}}{x^2+\frac{1}{2}}$$

$$= \sqrt{\pi}\,\text{erf}\,x + \frac{xe^{-x^2}}{x^2+\frac{1}{2}},$$

using the definition of erf x given in 5.11.

6. INTEGRAL INEQUALITIES

[First and second mean-value theorems, Cauchy–Schwarz inequality]

6.1 Show that $1 < \int_2^\infty \dfrac{2x^2}{x^4-1}\, dx < \log_e 3$, and obtain the exact value of the integral.

$$\frac{2}{x^2} < \frac{2x^2}{x^4-1} < \frac{2}{x^2-1} \quad \text{for } |x| > 1.$$

Hence

$$\int_2^\infty \frac{2}{x^2}\, dx < \int_2^\infty \frac{2x^2}{x^4-1}\, dx < \int_2^\infty \frac{2}{x^2-1}\, dx$$

or

$$1 < \int_2^\infty \frac{2x^2}{x^4-1}\, dx < \log_e 3.$$

Exact evaluation: partial fractions give

$$\frac{x^2}{x^4-1} = \frac{1}{2}\left\{\frac{1}{x^2-1} + \frac{1}{x^2+1}\right\}.$$

Hence integral is

$$-\frac{1}{2}\left[\log_e\left(\frac{x+1}{x-1}\right)\right]_2^\infty + \frac{\pi}{2} - \tan^{-1} 2 = \frac{\pi}{2} - \tan^{-1} 2 + \tfrac{1}{2}\log_e 3.$$

6.2 Show that

$$1 < \int_0^1 \frac{dx}{(2-x)\sqrt{1-x}} < 2.$$

For $0 < x < 1$, it is easily seen that

$$\frac{1}{2\sqrt{1-x}} < \frac{1}{(2-x)\sqrt{1-x}} < \frac{1}{\sqrt{1-x}}.$$

Since

$$\int_0^1 \frac{1}{\sqrt{1-x}}\, dx = \lim_{\varepsilon \to 1} \int_0^\varepsilon \frac{dx}{\sqrt{1-x}} = 2,$$

integrating each term of the inequality gives the required result.

6.3 By integrating the inequalities

$$\frac{1}{e\sqrt{x}} \leqslant \frac{e^{-x}}{\sqrt{x}} \leqslant \frac{1}{\sqrt{x}},$$

show that the improper integral $\int_0^1 \frac{e^{-x}}{\sqrt{x}}\, dx$ exists, and evaluate it approximately by substituting $x = t^2$ and expanding the integrand term-by-term.

$$\int_0^1 \frac{dx}{\sqrt{x}} = \lim_{\varepsilon \to 0} \int_\varepsilon^1 \frac{dx}{\sqrt{x}} = 2. \text{ Hence, on integrating,}$$

$$\frac{2}{e} < \int_0^1 \frac{e^{-x}}{\sqrt{x}}\, dx < 2.$$

If we write $x = t^2$, $\int_0^1 \frac{e^{-x}}{\sqrt{x}}\, dx$ becomes

$$2\int_0^1 e^{-t^2}\, dt = 2\int_0^1 \left(1 - t^2 + \frac{t^4}{2!} - \frac{t^6}{3!} + \cdots\right) dt$$

$$= 2\left(1 - \frac{1}{3\cdot 1!} + \frac{1}{5\cdot 2!} - \frac{1}{7\cdot 3!} + \cdots\right) \approx 0.913.$$

6.4 Evaluate the mean value of $f(x) = \dfrac{1}{4+x^2}$ over the range $0 \leqslant x \leqslant 2$, and verify the first mean-value theorem for integrals.

Basic result

The mean value of $f(x)$ over $a \leqslant x \leqslant b$ is

$$\frac{1}{(b-a)}\int_a^b f(x)\, dx.$$

Hence the mean value of $1/(4+x^2)$ in $0 \leqslant x \leqslant 2$ is

$$\frac{1}{2}\int_0^2 \frac{dx}{4+x^2} = \frac{1}{4}\left[\tan^{-1}\left(\frac{x}{2}\right)\right]_0^2 = \frac{\pi}{16}.$$

The first mean-value theorem states that there exists an x, say

$x = \xi$, in $a \le x \le b$ such that $f(\xi)$ is equal to the mean value. Hence $1/(4+\xi^2) = \pi/16$ gives $\xi = 1.04$, which lies in the range $(0, 2)$.

6.5 Use the first mean-value theorem to show that

$$\frac{\pi}{2} < \int_0^{\pi/2} \frac{dx}{\sqrt{1-\frac{1}{4}\sin^2 x}} < \frac{\pi}{\sqrt{3}}.$$

From the theorem (see 6.4) (taking $a = 0$, $b = \pi/2$)

$$\int_0^{\pi/2} \frac{dx}{\sqrt{1-\frac{1}{4}\sin^2 x}} = \frac{1}{\sqrt{1-\frac{1}{4}\sin^2 \xi}}\left(\frac{\pi}{2}-0\right), \quad \text{where} \quad 0 \le \xi \le \frac{\pi}{2}.$$

Putting $\xi = 0$ and $\pi/2$ (in turn) we find that the integral is bounded by the values $\pi/2$ and $\pi/\sqrt{3}$ as required.

6.6 The second mean-value theorem for integrals states that if $f(x)$ and $g(x)$ are continuous in $a \le x \le b$, and if $g(x)$ is positive in this range, then

$$\int_a^b f(x)g(x)\,dx = f(\xi)\int_a^b g(x)\,dx,$$

where $a \le \xi \le b$. Using this theorem show that if

$$I = \int_0^{\pi/2} e^{-x/15}\sin^2 x\,dx$$

then $0.90\,\dfrac{\pi}{4} \le I \le \dfrac{\pi}{4}$.

Let $f(x) = e^{-x/15}$ and $g(x) = \sin^2 x$. Then

$$I = e^{-\xi/15}\int_0^{\pi/2}\sin^2 x\,dx = \frac{\pi}{4}e^{-\xi/15},$$

where $0 \le \xi \le \pi/2$. Hence $0.90\,\dfrac{\pi}{4} \le I \le \dfrac{\pi}{4}$.

6.7 Verify the Cauchy–Schwarz inequality

$$\left|\int_a^b f(x)g(x)\,dx\right|^2 \le \left\{\int_a^b |f(x)|^2\,dx \int_a^b |g(x)|^2\,dx\right\}$$

for $f(x) = x^2$ and $g(x) = x-1$, where $a = 0$ and $b = 2$.

$$\left| \int_0^2 x^2(x-1)\,dx \right| = \left| \left(\frac{x^4}{4} - \frac{x^3}{3} \right)^2_0 \right| = \frac{4}{3}.$$

$$\int_0^2 |x^2|^2\,dx = \left(\frac{x^5}{5} \right)^2_0 = \frac{32}{5}.$$

$$\int_0^2 |x-1|^2\,dx = \int_0^1 (1-x)^2\,dx + \int_1^2 (x-1)^2\,dx = \frac{2}{3}.$$

(N.B. $|x-1| = 1-x$ for $x \le 1$.)

Since $\frac{32}{5} \cdot \frac{2}{3} > \left(\frac{4}{3} \right)^2$, the inequality is verified.

7. POWER SERIES AND CONVERGENCE

[Maclaurin and Taylor series, D'Alembert's ratio test, limits, Leibnitz–Maclaurin method, series expansion of integrals, approximate solution of $f(x) = 0$, iterative schemes, Newton–Raphson method]

7.1 Obtain the first three terms of the Maclaurin series for
(a) $\sqrt{1+x}$, (b) $\cos x$, (c) $\sec x$.

Maclaurin expansion:

$$f(x) = f(0) + xf'(0) + \frac{x^2}{2!} f''(0) + \cdots + \frac{x^r}{r!} f^{(r)}(0) + \cdots$$

(a) $f(x) = \sqrt{1+x}$, $f'(x) = \frac{1}{2}(1+x)^{-1/2}$, $f''(x) = -\frac{1}{4}(1+x)^{-3/2}$.
  $f(0) = 1$, $f'(0) = \frac{1}{2}$, $f''(0) = -\frac{1}{4}$.

Hence $f(x) = 1 + \dfrac{x}{2} - \dfrac{x^2}{8} + \cdots$

(b) $f(x) = \cos x$, $f'(x) = -\sin x$, $f''(x) = -\cos x$, and so on.
  $f(0) = 1$, $f'(0) = 0$, $f''(0) = -1$, $f'''(0) = 0$, $f''''(0) = 1$.

Hence $f(x) = 1 - \dfrac{x^2}{2!} + \dfrac{x^4}{4!} - \cdots$

(c) $f(x) = \sec x$,
  $f'(x) = \sec x \tan x$,
  $f''(x) = \sec^3 x + \sec x \tan^2 x$,
  $f'''(x) = 5 \sec^3 x \tan x + \sec x \tan^3 x$,
  $f''''(x) = 5 \sec^5 x + 15 \sec^3 x \tan^2 x + 3 \sec^3 x \tan^2 x + \sec x \tan^4 x$.
  $f(0) = 1$, $f'(0) = 0$, $f''(0) = 1$, $f'''(0) = 0$, $f''''(0) = 5$.

Hence $f(x) = 1 + \dfrac{x^2}{2!} + \dfrac{5x^4}{4!} + \cdots$

7.2 Show that there are no Maclaurin expansions for the functions $\operatorname{cosec} x$ and e^{-1/x^2}.

$\operatorname{cosec} x$ is undefined at $x = 0$, as are its derivatives. Hence no Maclaurin series exists.

Expressions of the form $e^{-1/x^2}/x^n$ $(n \geqslant 0)$ all tend to zero as $x \to 0$. Hence e^{-1/x^2} is not expressible as a Maclaurin series.

7.3 Use Maclaurin's theorem to derive the first two non-zero terms in the power series expansion of $\sin x$, $\tan^{-1} x$ and $\log_e (1+x)$ about $x = 0$. Hence calculate

$$\lim_{x \to 0} \left\{ \frac{\sin x - \tan^{-1} x}{x^2 \log_e (1+x)} \right\}. \tag{L.U.}$$

Taking the first few terms of the Maclaurin series we have $\sin x = x - \dfrac{x^3}{3!} + \cdots$, $\tan^{-1} x = x - \dfrac{x^3}{3} + \cdots$, $\log_e (1+x) = x - \dfrac{x^2}{2} + \cdots$

$$\frac{\sin x - \tan^{-1} x}{x^2 \log_e (1+x)} = \frac{(x - x^3/3! + \cdots) - (x - x^3/3 + \cdots)}{x^2(x - x^2/2 + \cdots)}$$

$$= \frac{x^3/6 + \cdots}{x^3 - x^4/2 + \cdots} \to \frac{1}{6} \quad \text{as } x \to 0.$$

7.4 Show that the Maclaurin series for $f(x) = \log_e \left(\dfrac{1+x}{1-x} \right)$ converges for $-1 < x < 1$, and that $f(x)$ may be represented by the series

$$2\left(x + \frac{x^3}{3} + \frac{x^5}{5} + \cdots \right).$$

The remainder after n terms is $R_{2n+1} = 2\left(\dfrac{x^{2n+1}}{2n+1} + \dfrac{x^{2n+3}}{2n+3} + \cdots \right)$.

Show that

$$|R_{2n+1}| < \frac{2}{2n+1} |x|^{2n+1} \left(\frac{1}{1-x^2} \right).$$

Basic result

D'Alembert's ratio test for convergence of a power series with the general form $f(x) = a_0 + a_1 x + a_2 x^2 + \cdots + a_n x^n + \cdots$ is

$$\lim_{n \to \infty} \left| \frac{a_{n+1} x^{n+1}}{a_n x^n} \right| \begin{cases} < 1, & \text{convergent,} \\ > 1, & \text{divergent.} \end{cases}$$

Maclaurin expansion of $\log_e(1+x) = x - \dfrac{x^2}{2} + \dfrac{x^3}{3} - \cdots + (-1)^n\dfrac{x^n}{n} + \cdots$.
By D'Alembert's test

$$\lim_{n\to\infty}\left|\frac{a_{n+1}x^{n+1}}{a_n x^n}\right| = \lim_{n\to\infty}\left|\frac{nx}{n+1}\right| = \lim_{n\to\infty}\left|\frac{n}{n+1}\right||x| = |x|;$$

hence convergent if $|x| < 1$.

Similarly $\log_e(1-x) = -x - \dfrac{x^2}{2} - \dfrac{x^3}{3} - \cdots - \dfrac{x^n}{n} - \cdots$ is convergent for $|x| < 1$.

Hence $\log_e\left(\dfrac{1+x}{1-x}\right) = \log_e(1+x) - \log_e(1-x) = 2\left(x + \dfrac{x^3}{3} + \dfrac{x^5}{5} + \cdots\right)$
is convergent for $-1 < x < 1$.

$$|R_{2n+1}| = 2|x^{2n+1}|\left|\left(\frac{1}{2n+1} + \frac{x^2}{2n+3} + \frac{x^4}{2n+5} + \cdots\right)\right|$$

$$< \frac{2}{2n+1}|x^{2n+1}|(1+x^2+x^4+\cdots)$$

$$= \frac{2}{2n+1}|x^{2n+1}|(1-x^2)^{-1},$$

since, for $|x| < 1$, $(1-x^2)^{-1} = 1 + x^2 + x^4 + \cdots$ by the binomial series.

7.5 Using D'Alembert's ratio test, investigate the convergence of the series

(a) $\displaystyle\sum_{n=1}^{\infty} \frac{(n+1)x^n}{n(n+2)}$, (b) $\displaystyle\sum_{n=1}^{\infty} \frac{x^n}{n^2}$.

(a) $a_n = \dfrac{n+1}{n(n+2)}$, $a_{n+1} = \dfrac{n+2}{(n+1)(n+3)}$.

$$\lim_{n\to\infty}\left|\frac{a_{n+1}x^{n+1}}{a_n x^n}\right| = \lim_{n\to\infty}\left|\frac{(n+2)x^{n+1}}{(n+1)(n+3)}\frac{n(n+2)}{(n+1)x^n}\right|$$

$$= \lim_{n\to\infty}\left\{\left|\frac{(n+2)^2 n}{(n+1)^2(n+3)}\right||x|\right\} = |x|.$$

Hence convergent for $|x| < 1$, and divergent for $|x| > 1$. At $x = \pm 1$ the ratio test does not give an answer, and the series must be examined separately at these points. At $x = 1$ the series becomes

$$\frac{2}{1\cdot3} + \frac{3}{2\cdot4} + \frac{4}{3\cdot5} + \cdots + \frac{(n+1)}{n(n+2)} + \cdots$$

The nth term may be expressed as partial fractions to give $1/2n + 1/2(n+2)$. Since $\sum_1^\infty 1/n$ (the harmonic series) is known to diverge, each term of the partial fraction is the nth term of a divergent series. Hence the series diverges at $x = 1$. At $x = -1$, the series becomes

$$-\frac{2}{1\cdot 3} + \frac{3}{2\cdot 4} - \frac{4}{3\cdot 5} + \cdots$$

This is an oscillating series whose terms decrease to zero. Hence convergent.

(b) $a_n = \dfrac{1}{n^2}$. Hence

$$\lim_{n\to\infty} \left| \frac{a_{n+1}x^{n+1}}{a_n x^n} \right| = \lim_{n\to\infty} \left| \frac{(n+1)^2 x}{n^2} \right|$$

$$= |x| \lim_{n\to\infty} \left| \left(\frac{n+1}{n}\right)^2 \right| = |x|.$$

Convergent if $|x| < 1$, divergent if $|x| > 1$. At $x = 1$ the series converges (using basic result that $\sum_1^\infty 1/n^2$ converges). At $x = -1$ series oscillates with terms decreasing to zero. Hence convergent.

7.6   Prove that if $y = e^{\sin^{-1} x}$, then $(1-x^2)\,d^2y/dx^2 - x\,dy/dx - y = 0$ and that

$$(1-x^2)\frac{d^{n+2}y}{dx^{n+2}} - (2n+1)x\frac{d^n y}{dx^n} - (n^2+1)\frac{d^n y}{dx^n} = 0.$$

Hence verify the Maclaurin expansion

$$e^{\sin^{-1} x} = 1 + x + \frac{x^2}{2} + \frac{x^3}{3} + \frac{5}{24}x^4 + \cdots$$

Taking logs: $\log_e y = \sin^{-1} x$, $(1/y)\,dy/dx = 1/\sqrt{1-x^2}$. Hence $dy/dx = y/\sqrt{1-x^2}$,

$$\frac{d^2y}{dx^2} = \frac{dy}{dx}\cdot\frac{1}{\sqrt{1-x^2}} + \frac{x}{(1-x^2)^{3/2}}\,y$$

and

$$(1-x^2)\frac{d^2y}{dx^2} = \sqrt{1-x^2}\,\frac{dy}{dx} + \frac{x}{\sqrt{1-x^2}}\,y.$$

But $x\dfrac{dy}{dx} = \dfrac{x}{\sqrt{1-x^2}}\,y$. Therefore

$$(1-x^2)\frac{d^2y}{dx^2} - x\frac{dy}{dx} = \sqrt{1-x^2}\,\frac{dy}{dx} + \frac{xy}{\sqrt{1-x^2}} - \frac{xy}{\sqrt{1-x^2}} = y,$$

giving the required result.

Now using Leibnitz's theorem (see 4.8) to differentiate each side n times we have

$$\left[(1-x^2)\frac{d^{n+2}y}{dx^{n+2}} + {}^nC_1(-2x)\frac{d^{n+1}y}{dx^{n+1}} - 2^nC_2\frac{d^ny}{dx^n}\right]$$

$$- \left[x\frac{d^{n+1}y}{dx^{n+1}} + {}^nC_1\frac{d^ny}{dx^n}\right] - \frac{d^ny}{dx^n} = 0.$$

Hence

$$(1-x^2)\frac{d^{n+2}y}{dx^{n+2}} - (2n+1)x\frac{d^{n+1}y}{dx^{n+1}} - (n^2+1)\frac{d}{dx^n} = 0.$$

At $x=0$,

$$\left(\frac{d^{n+2}y}{dx^{n+2}}\right)_{x=0} = (n^2+1)\left(\frac{d^ny}{dx^n}\right)_{x=0} \qquad \text{(recurrence relation)}.$$

Now since $y = e^{\sin^{-1}x}$, $y(0) = 1$, $y'(0) = 1$. Putting $n = 0$ in the recurrence relation gives $y''(0) = 1$; $n = 1, y'''(0) = 2y'(0) = 2$; $n = 2$, $y^{iv}(0) = 5$, and so on. Hence Maclaurin series is

$$1 + x + \frac{x^2}{2!} + \frac{2x^3}{3!} + \frac{5x^4}{4!} + \cdots$$

7.7 The functions $f(x)$, $g(x)$ and their first derivatives are all zero at $x = a$. By considering their Taylor expansions about $x = a$ show that

$$\lim_{x\to a}\left(\frac{f(x)}{g(x)}\right) = \frac{f''(a)}{g''(a)}.$$

Hence, or otherwise, evaluate

$$\lim_{x\to 0}\left\{\frac{[\log_e(1+x)]^2}{x\tan^{-1}x}\right\}. \qquad\qquad \text{(L.U.)}$$

$$f(x) = f(a) + (x-a)f'(a) + \frac{1}{2!}(x-a)^2f''(a) + \cdots \quad \text{(Taylor series) and}$$

likewise for $g(x)$. Hence if $f(a) = g(a) = 0$

$$\lim_{x \to a} \left(\frac{f(x)}{g(x)} \right) = \lim_{x \to a} \left\{ \frac{(x-a)f'(a) + \frac{1}{2}(x-a)^2 f''(a) + \cdots}{(x-a)g'(a) + \frac{1}{2}(x-a)^2 g''(a) + \cdots} \right\}$$

$$= \begin{cases} \dfrac{f'(a)}{g'(a)} & f'(a), \ g'(a) \text{ not both zero (l'Hôpital's rule)} \\[2ex] \dfrac{f''(a)}{g''(a)} & \text{if } f'(a) = g'(a) = 0 \text{ and } f''(a), \ g''(a) \text{ not} \\ & \text{both zero.} \end{cases}$$

Now write $f(x) = [\log_e (1+x)]^2$, which is zero, as is its first derivative $2[\log_e (1+x)]/(1+x)$ at $x = 0$. Likewise $g(x) = x \tan^{-1} x$ is zero at $x = 0$, as is its first derivative $x/(1+x^2) + \tan^{-1} x$. Differentiating once more,

$$f''(x) = \frac{2}{(1+x)^2} - \frac{2 \log_e (1+x)}{1+x},$$

whence $f''(0) = 2$. Similarly

$$g''(x) = \frac{1}{1+x^2} - \frac{2x^2}{(1+x)^2} + \frac{1}{1+x^2},$$

so $g''(0) = 2$. Hence, by the above result,

$$\text{limit} = \frac{f''(0)}{g''(0)} = 1.$$

7.8 Expand $f(x) = \sqrt{x}$ about $x = 1$ using Taylor's series, and show that the expansion is valid only for $0 \leqslant x \leqslant 2$.

$f(x) = \sqrt{x}$, $f(1) = 1$, $f'(1) = \frac{1}{2}$, $f''(1) = -\frac{1}{4}$, $f'''(1) = \frac{3}{8}$.
Hence the Taylor series about $x = 1$ is

$$1 + \frac{1}{2}(x-1) - \frac{1}{2!}\frac{1}{4}(x-1)^2 + \frac{1}{3!}\frac{3}{8}(x-1)^3 - \cdots$$

The nth term of this series is

$$\frac{1}{n!}(-1)^{n-1} \frac{1 \cdot 3 \cdot 5 \cdot 7 \cdots (2n-3)}{2^n}(x-1)^n.$$

By D'Alembert's ratio test (see 7.4)

$$\lim_{n \to \infty} \left| \frac{a_{n+1}(x-1)^{n+1}}{a_n(x-1)^n} \right| = \lim_{n \to \infty} \left| \frac{1}{2}\frac{2n-1}{n+1}(x-1) \right| = |x-1|.$$

Hence series is convergent if $|x-1| < 1$, that is, if $0 < x < 2$. When $x = 0$ the series is $1 - \frac{1}{2} - \frac{1}{2!}\frac{1}{4} - \frac{1}{3!}\frac{3}{8} - \cdots$, which converges, and when $x = 2$ series is $1 + \dfrac{1}{2} - \dfrac{1}{2!}\dfrac{1}{4} + \dfrac{1}{3!}\dfrac{3}{8} - \cdots$, which also converges. Hence given series is convergent for $0 \leqslant x \leqslant 2$.

7.9 Expand the integrand in series to show that

$$\frac{2}{\pi} \int_0^{\pi/2} \cos(x \sin t)\, dt = 1 - \frac{x^2}{4} + \frac{x^4}{64} - \cdots$$

(This function is the Bessel function of zero order, denoted by $J_0(x)$ – see also 17.1.)

$$\cos(x \sin t) = 1 - \frac{(x \sin t)^2}{2!} + \frac{(x \sin t)^4}{4!} - \cdots \quad \text{using the Maclaurin}$$

series expansion of the cosine function.
Hence

$$\frac{2}{\pi} \int_0^{\pi/2} \left(1 - \frac{(x \sin t)^2}{2!} + \frac{(x \sin t)^4}{4!} - \cdots \right) dt$$

$$= \frac{2}{\pi} \left[\frac{\pi}{2} - \frac{x^2}{2!}\frac{\pi}{4} + \frac{x^4}{4!}\frac{3\pi}{16} - \cdots \right]$$

$$= 1 - \frac{x^2}{2} + \frac{x^4}{64} - \cdots$$

using the standard integral

$$\int_0^{\pi/2} \sin^n t\, dt = \frac{(n-1)(n-3)\cdots 3 \cdot 1}{n(n-2)\cdots 4 \cdot 2}\frac{\pi}{2}$$

for n even $(n \neq 0)$.

7.10 Evaluate approximately by series expansion

(a) $\displaystyle\int_0^{\pi/2} \frac{\sin x}{x}\, dx$, (b) $\displaystyle\int_0^{0.2} \left\{ \frac{1}{\tan \sqrt{x}} - \frac{1}{\sqrt{x}} \right\} dx$.

(a) $\displaystyle\int_0^{\pi/2} \frac{\sin x}{x}\, dx = \int_0^{\pi/2} \frac{1}{x}\left(x - \frac{x^3}{3!} + \frac{x^5}{5!} - \cdots \right) dx$

$$= \int_0^{\pi/2} \left(1 - \frac{x^2}{3!} + \frac{x^4}{5!} - \cdots \right) dx$$

$$= \left(x - \frac{x^3}{18} + \frac{x^5}{600} - \frac{x^7}{35280} + \cdots \right)_0^{\pi/2}$$

$$= \frac{\pi}{2}\left(1 - \frac{\pi^2}{72} + \frac{\pi^4}{9600} - \cdots \right) = 1.371 \quad \text{to three terms.}$$

(b) Using the Maclaurin's series $\tan y = y + \dfrac{y^3}{3} + \dfrac{2}{15}y^5 + \cdots$, we have

$$\frac{1}{\tan\sqrt{x}} - \frac{1}{\sqrt{x}} = \left(\frac{1}{x^{1/2}+\frac{1}{3}x^{3/2}+\frac{2}{15}x^{5/2}+\cdots} - \frac{1}{x^{1/2}}\right)$$

$$= \frac{1}{\sqrt{x}}\left[\left(1+\frac{x}{3}+\frac{2x^2}{15}+\cdots\right)^{-1}-1\right]$$

$$= \frac{1}{\sqrt{x}}\left(1-\frac{x}{3}-\frac{x^2}{45}\ (+\,\text{terms in }x^3\text{ and higher})-1\right)$$

$$= \frac{1}{\sqrt{x}}\left(-\frac{x}{3}-\frac{x^2}{45}+\cdots\right)\quad\text{neglecting higher powers of }x.$$

Integrating this series gives

$$\int_0^{0.2}\left(\frac{1}{\tan\sqrt{x}}-\frac{1}{\sqrt{x}}\right)dx \simeq -0.198.$$

7.11 If $f(x)=0$ has a root at $x=x_0$ near a, show that

$$x_0 \simeq a - \frac{f(a)}{f'(a)}\qquad\text{(Newton's method)}.$$

Show that a better approximation to the root is

$$x_0 \simeq a - \frac{f(a)}{f'(a)} - \frac{[f(a)]^2 f''(a)}{2[f'(a)]^3}.$$

Evaluate these approximations to the root of $x^3+3x^2+6x-3=0$ which is near $x(=a)=\frac{1}{2}$.

Basic result

Taylor series is $f(x_0)=f(a)+(x_0-a)f'(a)+\dfrac{1}{2!}(x_0-a)^2 f''(a)+\cdots$

Hence since $f(x_0)=0$, then $x_0-a\simeq-f(a)/f'(a)$, neglecting terms in $(x_0-a)^2$ and higher powers. To obtain a better approximation we insert this value of x_0-a into the quadratic term of the Taylor series. Then

$$0 = f(a)+(x_0-a)f'(a)+\frac{1}{2}\frac{[f(a)]^2 f''(a)}{[f'(a)]^2},$$

whence $x_0-a\simeq-\dfrac{f(a)}{f'(a)}-\dfrac{[f(a)]^2 f''(a)}{2[f'(a)]^3}$, neglecting higher order terms.

If $f(x) = x^3 + 3x^2 + 6x - 3$ a crude approximation to the root is $a = \frac{1}{2}$.
Then

$$f(\tfrac{1}{2}) = \tfrac{7}{8}, \qquad f'(\tfrac{1}{2}) = \tfrac{39}{4}, \qquad f''(\tfrac{1}{2}) = 9.$$

Hence $x_0 - \frac{1}{2} \simeq -\frac{7}{8}/\frac{39}{4} = -\frac{7}{78}$ so $x_0 = 0.410$ is the first approximation,
and

$$x_0 - \frac{1}{2} \simeq -\frac{7}{78} - \frac{1}{2}\frac{(\tfrac{7}{8})^2 \cdot 9}{(\tfrac{39}{4})^3},$$

or $x_0 = 0.406$, is a better approximation.

7.12 (a) Show that

$$x_{n+1} = \frac{1}{2}\left(x_n + \frac{N}{x_n}\right),$$

where $n = 1, 2, 3, \ldots$ is a suitable iterative scheme for $\sqrt{N}$. Hence
evaluate $\sqrt{137}$ given the first approximation $x_1 = 10$.
(b) Show that

$$x_{n+1} = x_n(2 - Nx_n)$$

is a suitable iterative scheme for $1/N$. Evaluate $\frac{1}{3}$ given $x_1 = \frac{1}{10}$.

(a) As $n \to \infty$, both x_n, x_{n+1} tend to a limit l, say, if the sequence is
convergent. Therefore $l = \frac{1}{2}(l + (N/l))$, whence $l = \sqrt{N}$.

$$x_2 = \frac{1}{2}\left(10 + \frac{137}{10}\right) = 11.85$$

$$x_3 = \frac{1}{2}\left(11.85 + \frac{137}{11.85}\right) = 11.706$$

$$x_4 = \frac{1}{2}\left(11.706 + \frac{137}{11.706}\right) = 11.705$$

$\sqrt{137} = 11.705$ (to 3 decimal places).

(b) If l is the limit of the sequence, then

$$l = l(2 - Nl), \quad \text{whence} \quad l = \frac{1}{N}.$$

$x_2 = \frac{1}{10}(2 - 0.3) = 0.17,$
$x_3 = 0.17(2 - 0.51) = 0.246,$
$x_4 = 0.246(2 - 0.738) = 0.310,$
$x_5 = 0.310(2 - 0.930) = 0.332,$
$x_6 = 0.332(2 - 0.996) = 0.333$ (to 3 decimal places).

7.13 Use the Newton–Raphson iterative scheme

$$x_{n+1} = x_n - \frac{f(x_n)}{f'(x_n)}$$

to find the smallest root of $f(x) = x \sin x - 1 = 0$ taking $x_1 = 1$.

$f(x) = x \sin x - 1$, $f'(x) = x \cos x + \sin x$.
$f(1) = \sin 1 - 1 = 0.841 - 1 = -0.159$,
$f'(1) = \cos 1 + \sin 1 = 1.382$.
Hence $x_2 = 1 + \dfrac{0.159}{1.382} = 1.115$.
$f(1.115) = 1.115 \sin 1.115 - 1 = 0.001$,
$f'(1.115) = \cos 1.115 + \sin 1.115 = 1.338$.

$$x_3 = 1.115 - \frac{0.001}{1.338} = 1.114.$$

This is correct to three decimal places.

7.14 If $T_{n+1} = a^{T_n}$, where $a > 0$ and $n = 1, 2, 3, \ldots$ show that $T_4 = a^{a^{a^{T_1}}}$. If $T_1 = 1$, verify numerically that this sequence converges for large n if

$$e^{-e}(= 0.065988) \leqslant a \leqslant e^{1/e}(= 1.44467).$$

Evaluate the limiting values of this sequence at the end points of this range.

As a typical value take $a = 1.2$. Then numerical evaluation gives

$$1.2^{1.2} = 1.244, \quad 1.2^{1.244} = 1.254, \quad 1.2^{1.254} = 1.257,$$
$$1.2^{1.257} = 1.258, \quad 1.2^{1.258} = 1.258,$$

and so on. By taking values of a close to 0.06599 and 1.4446 the sequence is seen to converge less rapidly. For values of a outside the given range the numerical values of $a^{a^{a^{\cdots}}}$ are easily seen to increase without limit.

Assuming the range of convergence (as established analytically by Euler) we have for the limiting value l of the sequence (since $\lim_{n \to \infty} T_{n+1} = \lim_{n \to \infty} T_n = l$)

$$l = a^l,$$

which if $a = e^{1/e}$ gives $l = e$. If $a = e^{-e}$, then $l = e^{-el}$, whence $l = 1/e$.

7.15 Show that $x^n + y^n = z^n$, where x, y, z are *consecutive* integers $2n-1$, $2n$, $2n+1$, does not have solutions if n is a positive integer

>2. (This problem relates to Fermat's unproven conjecture that there are no positive integer solutions *at all* for integers $n > 2$.)

Consider $(2n-1)^n + (2n)^n - (2n+1)^n = R$ (say).
Then

$$\left(1-\frac{1}{2n}\right)^n + 1 - \left(1+\frac{1}{2n}\right)^n = \frac{R}{(2n)^n},$$

whence, using the binomial theorem,

$$\left(1 - \frac{n}{2n} + \frac{n(n-1)}{2!}\left(\frac{1}{2n}\right)^2 - \frac{n(n-1)(n-2)(n-3)}{3!}\left(\frac{1}{2n}\right)^3 + \cdots\right) + 1$$

$$- \left(1 + \frac{n}{2n} + \frac{n(n-1)}{2!}\left(\frac{1}{2n}\right)^2 - \cdots\right)$$

$$= -\frac{(n-1)(n-2)}{24n^2} - \frac{(n-1)(n-2)(n-3)(n-4)}{1920n^4} - \cdots - \cdots$$

Each term in this infinite series is strictly negative except when $n = 1$ or 2. Hence if $n > 2$, $R < 0$ and no integer solutions exist. N.B. If $n = 1$, $x = 1$, $y = 2$, then $z = 3$ and if $n = 2$, $x = 3$, $y = 4$, then $z = 5$.

8. COMPLEX VARIABLE

[Modulus and argument, loci in Argand plane, roots of equations, integration, summation of series, complex functions, analytic functions, Cauchy–Riemann relations, conformal transformation]

8.1 Express in the form $x + iy$: (a) $(1+i)/(1-i)$, (b) $1/i^5$, (c) $(-\frac{1}{2}+i\sqrt{3}/2)^2$, (d) $i^{1/2}$.

(a) $\dfrac{1+i}{1-i} = \dfrac{(1+i)^2}{(1-i)(1+i)} = \dfrac{(1+i)^2}{1-i^2} = \dfrac{1}{2} \cdot 2i = i.$

(b) $\dfrac{1}{i^5} = \dfrac{1}{i}\dfrac{1}{i^4} = -i.$

(c) $\left(-\dfrac{1}{2}+i\dfrac{\sqrt{3}}{2}\right)^2 = \dfrac{1}{4} - \dfrac{3}{4} - \dfrac{2i\sqrt{3}}{4} = -\dfrac{1}{2} - i\dfrac{\sqrt{3}}{2}.$

(d) $i^{1/2} = x + iy$, $i = (x+iy)^2 = x^2 - y^2 + 2ixy$. Hence $x^2 - y^2 = 0$, $2xy = 1$.

$x = y = \pm\dfrac{1}{\sqrt{2}}$. Therefore $i^{1/2} = \dfrac{1}{\sqrt{2}}(1+i)$ and $-\dfrac{1}{\sqrt{2}}(1+i)$.

8.2 Prove that

$$\frac{1+\cos\theta+i\sin\theta}{1-\cos\theta+i\sin\theta} = \cot\frac{\theta}{2}\, e^{i(\theta-\pi/2)}.$$

Use half-angle formulae:

$$1+\cos\theta = 2\cos^2\frac{\theta}{2}, \quad 1-\cos\theta = 2\sin^2\frac{\theta}{2}, \quad \sin\theta = 2\sin\frac{\theta}{2}\cos\frac{\theta}{2}.$$

Then

$$\frac{1+\cos\theta+i\sin\theta}{1-\cos\theta+i\sin\theta} = \frac{2\cos^2\dfrac{\theta}{2}+2i\sin\dfrac{\theta}{2}\cos\dfrac{\theta}{2}}{2\sin^2\dfrac{\theta}{2}+2i\sin\dfrac{\theta}{2}\cos\dfrac{\theta}{2}}$$

$$= \cot\frac{\theta}{2}\,\frac{\cos\dfrac{\theta}{2}+i\sin\dfrac{\theta}{2}}{\sin\dfrac{\theta}{2}+i\cos\dfrac{\theta}{2}}.$$

Euler's formula: $e^{i\theta} = \cos\theta + i\sin\theta$ (prove by expanding $e^{i\theta}$, $\cos\theta$ and $\sin\theta$ as Maclaurin series and comparing real and imaginary parts). Hence

$$e^{i\theta/2} = \cos\frac{\theta}{2}+i\sin\frac{\theta}{2} \quad \text{and} \quad e^{i(\pi/2-\theta/2)} = \sin\frac{\theta}{2}+i\cos\frac{\theta}{2},$$

and result follows.

8.3 Find the moduli and arguments of the following:

(a) $1-i\sqrt{3}$, (b) $e^{i\pi/2}+\sqrt{2}\,e^{i\pi/4}$, (c) $(1+i)e^{i\pi/6}$.

Basic result

$z = x+iy = re^{i(\theta+2\pi n)}$, $n = 0, 1, 2, 3, \ldots$, where $r = |z|$ is the modulus of z and θ is its argument $-$ arg z.

(a) $z = 1-i\sqrt{3}$; modulus $(|z|) = \sqrt{1^2+(\sqrt{3})^2} = 2$, argument (arg z) $= \tan^{-1}(-\sqrt{3})$.

(b) $e^{i\pi/2} + \sqrt{2}\, e^{i\pi/4} = i + \sqrt{2}\left(\cos\dfrac{\pi}{4} + i\sin\dfrac{\pi}{4}\right)$

$$= i + \sqrt{2}\left(\dfrac{1}{\sqrt{2}} + \dfrac{i}{\sqrt{2}}\right) = 1 + 2i,$$

using Euler's formula (see 8.2)).

$\text{Modulus} = \sqrt{1^2 + 2^2} = \sqrt{5}, \quad \text{argument} = \tan^{-1} 2.$

(c) $(1+i)e^{i\pi/6} = (1+i)\left(\cos\dfrac{\pi}{6} + i\sin\dfrac{\pi}{6}\right) = (1+i)\left(\dfrac{\sqrt{3}}{2} + i\dfrac{1}{2}\right)$

$$= \left(\dfrac{\sqrt{3}-1}{2}\right) + i\left(\dfrac{\sqrt{3}+1}{2}\right).$$

$\text{Modulus} = \sqrt{\left(\dfrac{\sqrt{3}-1}{2}\right)^2 + \left(\dfrac{\sqrt{3}+1}{2}\right)^2} = \sqrt{2},$

$\text{argument} = \tan^{-1}\left(\dfrac{\sqrt{3}+1}{\sqrt{3}-1}\right).$

8.4 What is wrong with the following statement

$$1 = \sqrt{(-1)(-1)} = \sqrt{-1}\,\sqrt{-1} = i \cdot i = -1?$$

Every complex number has a modulus and argument. In any equation these quantities must agree on both sides. Now $1 = 1e^{2\pi i}$. But $\sqrt{-1}$ has two values: $1e^{\pi i/2}\ (= i)$ and $1e^{3\pi i/2}\ (= -i)$. The second value must be used for one of the $\sqrt{-1}$s to ensure that the arguments are the same on both sides. Hence

$$1 = \sqrt{(-1)(-1)} = \sqrt{-1}\,\sqrt{-1} = e^{\pi i/2}e^{3\pi i/2} = e^{2\pi i} = 1.$$

8.5 What plane curves are represented by the equations

(a) $|z - 1| = 2$, (b) $|z + 1| = |z - i|$, (c) $\text{Re}\,(z^2) = 1$,

where Re denotes the real part, (d) $z^2 - z\bar{z} + \bar{z}^2 = 0$ ($\bar{z} = x - iy$ is the complex conjugate of z).

(a) $|z - 1| = \sqrt{(x - 1 - iy)(x - 1 + iy)} = \sqrt{(x - 1)^2 + y^2} = 2.$

Hence $(x - 1)^2 + y^2 = 4$, which represents a circle centre 1, radius 2. (See Fig. 12a.)

(b) $|z + 1| = \sqrt{(x + 1)^2 + y^2}, \quad |z - i| = \sqrt{(x + i(y - 1))(x - i(y - 1))}$
$$= \sqrt{x^2 + (y - 1)^2}.$$

Hence $x^2 + y^2 + 2x + 1 = x^2 + y^2 - 2y + 1$, giving $y = -x$, which is a straight line (see Fig. 12b).

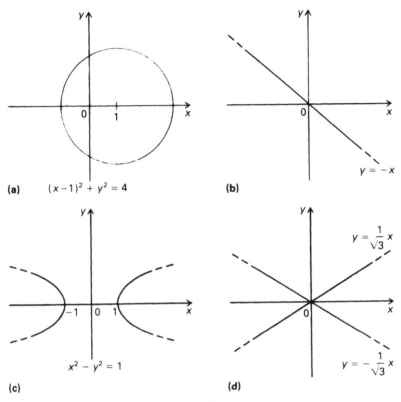

(a) $(x-1)^2 + y^2 = 4$

(b) $y = -x$

(c) $x^2 - y^2 = 1$

(d) $y = \dfrac{1}{\sqrt{3}}x$ $y = -\dfrac{1}{\sqrt{3}}x$

Fig. 12

(c) $\operatorname{Re}(z^2) = 1$; $z = x + iy$, $z^2 = x^2 + 2ixy - y^2$, $\operatorname{Re}(z^2) = x^2 - y^2$.

Hence $x^2 - y^2 = 1$, which is a hyperbola (see Fig. 12c).

(d) $z^2 - z\bar{z} + \bar{z}^2 = 0$; $(x+iy)^2 - (x+iy)(x-iy) + (x-iy)^2 = 0$.

Hence $x^2 - 3y^2 = 0$, which represents a pair of straight lines $y = \pm x/\sqrt{3}$ (see Fig. 12d).

8.6   Given that $C = \int e^{ax} \cos bx \, dx$ and $S = \int e^{ax} \sin bx \, dx$ (a, b are real constants), show that $C + iS = \int e^{(a+ib)x} \, dx$, and hence that

$$C = \frac{e^{ax}}{a^2 + b^2}(a \cos bx + b \sin bx),$$

$$S = \frac{e^{ax}}{a^2 + b^2}(a \sin bx - b \cos bx)$$

(integration constants omitted).

$$C + iS = \int e^{ax}(\cos bx + i \sin bx)\, dx$$

$$= \int e^{ax} e^{ibx}\, dx = \int e^{(a+ib)x}\, dx.$$

Integrating:

$$C + iS = \frac{e^{(a+ib)x}}{a + ib} = \frac{(a - ib)e^{(a+ib)x}}{a^2 + b^2} = \frac{(a - ib)}{a^2 + b^2}(\cos bx + i \sin bx)e^{ax}.$$

Comparing real and imaginary parts of each side

$$C = \frac{e^{ax}}{a^2 + b^2}(a \cos bx + b \sin bx), \quad S = \frac{e^{ax}}{a^2 + b^2}(a \sin bx - b \cos bx).$$

8.7 Prove that if $y = e^x \sin x$, then $d^n y/dx^n = 2^{n/2} e^x \sin(x + n\pi/4)$. Hence, or otherwise, calculate the first five terms of the Taylor series of y expanded about $x = \pi/2$. (L.U.)

$y = e^x \sin x = \operatorname{Im}\{e^{(1+i)x}\}$ (Im denotes imaginary part).

$$\frac{d^n y}{dx^n} = \operatorname{Im}\{(1+i)^n e^{(1+i)x}\} = \operatorname{Im}\{2^{n/2} e^{in\pi/4} e^{(1+i)x}\}$$

$$= 2^{n/2} e^x \operatorname{Im}\{e^{i(x+n\pi/4)}\} = 2^{n/2} e^x \sin\left(x + \frac{n\pi}{4}\right).$$

$$y(x) = y\left(\frac{\pi}{2}\right) + \left(x - \frac{\pi}{2}\right)y'\left(\frac{\pi}{2}\right) + \frac{\left(x - \frac{\pi}{2}\right)^2}{2!}y''\left(\frac{\pi}{2}\right) + \cdots$$

$$= e^{\pi/2} + \left(x - \frac{\pi}{2}\right)e^{\pi/2}\sqrt{2}\sin\left(\frac{\pi}{2} + \frac{\pi}{4}\right)$$

$$+ \frac{1}{2}\left(x - \frac{\pi}{2}\right)^2 e^{\pi/2} \cdot 2 \sin\left(\frac{\pi}{2} + \frac{\pi}{2}\right)$$

$$+ \frac{(x - \pi/2)^3}{3!} e^{\pi/2} 2^{3/2} \sin\left(\frac{\pi}{2} + \frac{3\pi}{4}\right) + \cdots$$

$$= e^{\pi/2}\left[1 + \left(x - \frac{\pi}{2}\right) - \frac{1}{3}\left(x - \frac{\pi}{2}\right)^3 - \frac{1}{6}\left(x - \frac{\pi}{2}\right)^4 - \frac{1}{30}\left(x - \frac{\pi}{2}\right)^5 + \cdots\right].$$

8.8 Solve the equation $z^6 + z^3 + 1 = 0$.

Let $z^3 = \omega$. Hence $\omega^2 + \omega + 1 = 0$ or $\omega = (-1 \pm \sqrt{-3})/2 = -\frac{1}{2} \pm i\sqrt{3}/2$. If $\omega = -\frac{1}{2} + i\sqrt{3}/2 = e^{i(2\pi/3 + 2\pi k)}$, where k is an integer, then, using De

Moivre's relation

$$z = \omega^{1/3} = [e^{i(2\pi/3 + 2\pi k)}]^{1/3} = e^{i(2\pi/9 + 2\pi k/3)}, \qquad \text{where } k = 0, 1, 2.$$

(N.B. other integral values of k simply reproduce the results of $k = 0, 1, 2$.) When

$$\omega = -\frac{1}{2} - i\frac{\sqrt{3}}{2}, \quad z = \omega^{1/3} = e^{i(4\pi/9 + 2\pi k/3)}, \qquad k = 0, 1, 2.$$

Both results may be combined in the form

$$z = e^{i(2\pi n/9)}, \qquad n = 1, 2, 4, 5, 7, 8.$$

8.9 Sum the series

$$S = \frac{\sin \theta}{\cos \theta} + \frac{\sin 2\theta}{\cos^2 \theta} + \frac{\sin 3\theta}{\cos^3 \theta} + \cdots + \frac{\sin n\theta}{\cos^n \theta}.$$

Let $C = \dfrac{\cos \theta}{\cos \theta} + \dfrac{\cos 2\theta}{\cos^2 \theta} + \dfrac{\cos 3\theta}{\cos^3 \theta} + \cdots + \dfrac{\cos n\theta}{\cos^n \theta}.$

Then

$$C + iS = \frac{e^{i\theta}}{\cos \theta} + \frac{e^{2i\theta}}{\cos^2 \theta} + \frac{e^{3i\theta}}{\cos^3 \theta} + \cdots + \frac{e^{ni\theta}}{\cos^n \theta}$$

$$= 1 + \left(\frac{e^{i\theta}}{\cos \theta}\right) + \left(\frac{e^{i\theta}}{\cos \theta}\right)^2 + \cdots + \left(\frac{e^{i\theta}}{\cos \theta}\right)^n - 1$$

$$= \left\{\left[1 - \left(\frac{e^{i\theta}}{\cos \theta}\right)^{n+1}\right] \Big/ \left(1 - \frac{e^{i\theta}}{\cos \theta}\right)\right\} - 1,$$

using the formula for the sum of a geometrical series with common ratio $e^{i\theta}/\cos \theta$. Taking the imaginary part we find ultimately

$$S = \frac{\cos^{n+1} \theta - \cos (n+1)\theta}{\cos^n \theta \sin \theta}.$$

8.10 Show that e^z has no zeros, and locate the zeros of $\cosh z$.

$e^z = e^x e^{iy}$. For $e^z = 0$, $e^x e^{iy} = 0$. Now since $e^x \neq 0$, $e^{iy} = 0$. Hence $\cos y + i \sin y = 0$ and therefore $\cos y = 0$, $\sin y = 0$ simultaneously, which is impossible. Hence $e^{iy} \neq 0$, and consequently $e^z \neq 0$ for all z.

$\cosh z = \dfrac{e^z + e^{-z}}{2}$. Therefore $e^z + e^{-z} = 0$ or $e^{2z} + 1 = 0$. Hence $e^{2z} = -1 = e^{\pi i + 2k\pi i}$, where $k = 0, 1, 2, \ldots$ and therefore $z = i\dfrac{\pi}{2}(1 + 2k)$.

8.11 Determine all possible values of (a) $\log_e (1+i)$, (b) $\sin^{-1} 2$, (c) $\tan^{-1} 2i$.

(a) $\log_e (1+i) = \log_e |1+i| + i \arg (1+i)$

$$= \log_e \sqrt{2} + i\left(\frac{\pi}{4} + 2k\pi\right), \qquad k = 0, 1, 2, \ldots.$$

(b) Let $\omega = \sin^{-1} z$, so that $z = \sin \omega = (e^{i\omega} - e^{-i\omega})/2i$ (using Euler's formula – see 8.2). Hence $e^{2i\omega} - 2ize^{i\omega} - 1 = 0$ (a quadratic in $e^{i\omega}$) or $e^{i\omega} = iz \pm \sqrt{1-z^2}$ (by solving the quadratic). Finally therefore

$$\omega = \frac{1}{i} \log_e \{iz \pm \sqrt{1-z^2}\}.$$

Now put $z = 2$.

$$\sin^{-1} 2 = \frac{1}{i} \log_e \{2i \pm \sqrt{-3}\} = \frac{1}{i} \log_e \{(2 \pm \sqrt{3})i\}$$

$$= \frac{1}{i}\left[\log_e (2 \pm \sqrt{3}) + i\left(\frac{\pi}{2} + 2k\pi\right)\right], \qquad k = 0, 1, 2, \ldots.$$

since $|(2 \pm \sqrt{3})i| = 2 \pm \sqrt{3}$.

Hence $\sin^{-1} 2 = \left(\frac{\pi}{2} + 2k\pi\right) - i \log_e (2 \pm \sqrt{3})$.

But $\log_e (2 - \sqrt{3}) = -\log_e (2 + \sqrt{3})$

(since $\log_e (2 - \sqrt{3}) + \log_e (2 + \sqrt{3}) = \log_e (2 - \sqrt{3})(2 + \sqrt{3}) = \log_e 1 = 0$).

Finally then $\sin^{-1} 2 = \frac{\pi}{2} + 2k\pi \pm i \log_e (2 + \sqrt{3})$.

(c) Let $\omega = \tan^{-1} z$. Then

$$z = \tan \omega = \frac{e^{i\omega} - e^{-i\omega}}{i(e^{i\omega} + e^{-i\omega})} = \frac{e^{2i\omega} - 1}{i(e^{2i\omega} + 1)}.$$

Hence $ize^{2i\omega} + iz = e^{2i\omega} - 1$ and therefore $e^{i\omega} = \pm\sqrt{(1+iz)/(1-iz)}$, giving

$$\omega = \frac{1}{i} \log_e \left(\pm\sqrt{\frac{1+iz}{1-iz}}\right).$$

Put $z = 2i$.

$$\omega = \tan^{-1} 2i = \frac{1}{i} \log_e \left(\pm\sqrt{-\frac{1}{3}}\right) = \frac{1}{i} \log_e \left(\pm\frac{i}{\sqrt{3}}\right).$$

Taking the positive sign:

$$\omega = \frac{1}{i} \log_e \left(\frac{i}{\sqrt{3}}\right) = \frac{1}{i}\left[\log_e \left|\frac{1}{\sqrt{3}}\right| + i\left(\frac{\pi}{2} + 2k\pi\right)\right]$$

$$= \frac{\pi}{2} + 2k\pi + i \log_e |\sqrt{3}|, \qquad k = 0, 1, 2, \ldots.$$

Taking the negative sign:

$$\omega = \frac{1}{i}\log_e\left(-\frac{i}{\sqrt{3}}\right) = \frac{1}{i}\left[\log_e\left|\frac{1}{\sqrt{3}}\right| + i\left(\frac{3\pi}{2} + 2k\pi\right)\right]$$

$$= \frac{3\pi}{2} + 2k\pi + i\log_e|\sqrt{3}|, \qquad k = 0, 1, 2, \ldots .$$

Hence both cases together give

$$\omega = \tan^{-1} 2i = \frac{\pi}{2}(1+2k) + i\log_e|\sqrt{3}|, \qquad k = 0, 1, 2, \ldots$$

8.12 Find the points, if any, at which the following functions are not analytic:

(a) $\dfrac{1}{(z+1)^3}$, (b) $\dfrac{\cos z}{z^2+1}$, (c) e^z.

(a) Function not defined at $z = -1$. Hence not analytic at this point. Analytic elsewhere.
(b) Function not defined when $z^2+1 = 0$ i.e. at $z = \pm i$. Hence not analytic at these two points. Analytic elsewhere.
(c) $e^z = e^{x+iy} = e^x(\cos y + i\sin y)$. Hence $u = e^x\cos y$, $v = e^x\sin y$ and the Cauchy–Riemann equations $\partial u/\partial x = \partial v/\partial y$, $\partial v/\partial x = -\partial u/\partial y$ are satisfied for all z. Hence, since the derivatives are continuous, e^z is analytic for all z.

8.13 Determine which of the following functions satisfy the Cauchy–Riemann equations:
(a) $z^3 - iz^2 + 1$, (b) $|z|^2$.

(a) $f(z) = u + iv = (x+iy)^3 - i(x+iy)^2 + 1$
$= (x^3 + 3ix^2y - 3xy^2 - iy^3) - i(x^2 + 2ixy - y^2) + 1$
$= (x^3 - 3xy^2 + 2xy + 1) + i(3x^2y - y^3 - x^2 + y^2).$

$\dfrac{\partial u}{\partial x} = 3x^2 - 3y^2 + 2y,$ $\dfrac{\partial v}{\partial y} = 3x^2 - 3y^2 + 2y.$ Hence $\dfrac{\partial u}{\partial x} = \dfrac{\partial v}{\partial y}.$

$\dfrac{\partial u}{\partial y} = -6xy + 2x,$ $\dfrac{\partial v}{\partial x} = 6xy - 2x.$ Hence $\dfrac{\partial u}{\partial y} = -\dfrac{\partial v}{\partial x}.$

Function satisfies the Cauchy–Riemann equations.
(b) $f(z) = |z|^2 = x^2 + y^2$. Hence $u = x^2 + y^2$, $v = 0$, and the Cauchy–Riemann equations are therefore not satisfied.

8.14 Find the constant c such that $e^{cx}\cos 3y$ is the real part of an analytic function, and find this function.

If $u = e^{cx} \cos 3y$, then $\dfrac{\partial u}{\partial x} = ce^{cx} \cos 3y = \dfrac{\partial v}{\partial y}$. Hence integrating with respect to y gives $v = \tfrac{1}{3}ce^{cx} \sin 3y + \theta(x)$, where $\theta(x)$ is an arbitrary function of integration. Similarly the other Cauchy–Riemann relation gives

$$\frac{\partial u}{\partial y} = -3e^{cx} \sin 3y = -\frac{\partial v}{\partial x}.$$

Hence integrating with respect to x we have $v = \dfrac{3}{c} e^{cx} \sin 3y + \phi(y)$, where $\phi(y)$ is an arbitrary function of integration.

Equating the two expressions for v gives

$$\tfrac{1}{3}ce^{cx} \sin 3y + \theta(x) = \frac{3}{c} e^{cx} \sin 3y + \phi(y),$$

whence $c^2 = 9$ or $c = \pm 3$, $\theta(x) = \phi(y) = k$ (arbitrary *constant*), and $f(z) = u + iv = e^{\pm 3z} + k$.

8.15 Find the fixed points of the transformation

$$\omega = 1 - \frac{1}{z-1}.$$

Fixed points defined by $\omega = z$. Hence $(z-1)^2 + 1 = 0$, which gives $z = 1 \pm i$.

8.16 Find the bilinear transformation which maps the points $z = 1$ to $\omega = 0$, $z = i$ to $\omega = 1$, and $z = -1$ to $\omega \to \infty$.

Bilinear transformation is of the form $\omega = \dfrac{\alpha z + \beta}{z + \gamma}$, where α, β and γ are constants.

$z = -1 \to \omega = \infty$; hence $\gamma = 1$. $z = 1 \to \omega = 0$; therefore $\alpha = -\beta$, and hence $\omega = \dfrac{\alpha(z-1)}{z+1}$. $z = i \to \omega = 1$; therefore $1 = \dfrac{\alpha(i-1)}{i+1}$ or $\alpha = -i$.

Hence finally $\omega = -i\left(\dfrac{z-1}{z+1}\right)$.

8.17 Find the region of the ω-plane into which the transformation $\omega = z^2$ maps the region bounded by $x = 1$, $y = 1$ and $x + y = 1$. Show that the mapping is conformal in this region.

$\omega = u + iv = z^2 = (x + iy)^2 = x^2 - y^2 + 2ixy$.
Hence $u = x^2 - y^2$, $v = 2xy$. Consider the boundary BC for which

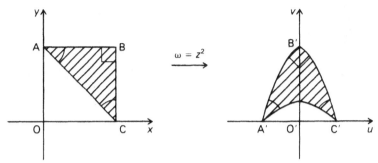

Fig. 13

$x = 1$ (see Fig. 13). Then $u = 1 - y^2$, $v = 2y$. Hence eliminating y we obtain $u = 1 - v^2/4$, which is a parabola B′C′ in the u, v plane, where $C' = (1, 0)$, $B' = (0, 2)$. Now on AB, $y = 1$. Hence $u = x^2 - 1$, $v = 2x$, and therefore $u = v^2/4 - 1$, which is a parabola passing through the points $A' = (-1, 0)$, $B' = (0, 2)$.

Finally the boundary $x + y = 1$ gives $u = 1 - 2y$, $v = 2y(1 - y)$. Hence $u^2 = 1 - 2v$ and AC becomes the parabola A′C′. Since $\omega = f(z) = z^2$ is an analytic function, and $f'(z) = 2z \neq 0$ within the specified region, the transformation is conformal and the angles at A, B, C are preserved at A′, B′ and C′.

8.18 Given that $\omega = \sin(z^2)$, find the equation of the curve in the ω-plane corresponding to the hyperbola $xy = c$ (c constant) in the z-plane.

$$\omega = u + iv = \sin[(x + iy)^2] = \sin(x^2 - y^2 + 2ixy)$$
$$= \sin(x^2 - y^2)\cos 2ixy + \cos(x^2 - y^2)\sin 2ixy.$$

Hence $u = \sin(x^2 - y^2)\cosh 2xy$, $v = \cos(x^2 - y^2)\sinh 2xy$.

$$xy = c \rightarrow \left(\frac{u}{\cosh 2c}\right)^2 + \left(\frac{v}{\sinh 2c}\right)^2 = 1, \quad \text{which is an ellipse.}$$

8.19 Show that the transformation $\omega = z + 1/z$ maps points on a circle $|z| = c$ ($\neq 1$) in the z-plane onto an ellipse in the ω-plane.

Let $z = re^{i\theta}$. Then $\omega = u + iv = (r + 1/r)\cos\theta + i(r - 1/r)\sin\theta$. On $r = |z| = c$, $u = (c + 1/c)\cos\theta$, $v = (c - 1/c)\sin\theta$. Therefore

$$\frac{u^2}{(c + 1/c)^2} + \frac{v^2}{(c - 1/c)^2} = 1.$$

8.20 Show that the transformation $\omega = z^2 + a^4/z^2$ maps the region $x^2 + y^2 \geq a^2$, $x \geq 0$, $y \geq 0$ in the z-plane into the upper half of the ω-plane. At what points is the transformation not conformal?

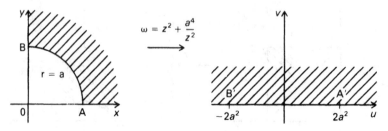

Fig. 14

Let $z = re^{i\theta}$. Then $u + iv = r^2 e^{2i\theta} + (a^4/r^2)e^{-2i\theta}$.

Therefore $u = \left(r^2 + \dfrac{a^4}{r^2}\right)\cos 2\theta$ and $v = \left(r^2 - \dfrac{a^4}{r^2}\right)\sin 2\theta$.

Now the region $x \geq 0$, $y \geq 0$ corresponds to $0 \leq \theta \leq \pi/2$. For $r^2 \geq a^2$ we see that $v \geq 0$, and that u may be positive or negative. Hence the mapping is into the upper half of the ω-plane. From $\omega = f(z) = z^2 + a^2/z^2$ we have $f'(z) = 2z - 2a^4/z^3$, which is zero at $z^4 = a^4$ and therefore gives the points

$$z = a, \ ae^{i\pi/2}, \ ae^{i\pi}, \ ae^{i3\pi/2}.$$

Only the first two of these points lie within the given region and, since $f'(z) = 0$, the transformation is not conformal at these points; consequently angles are not preserved (see Fig. 14).

9. HYPERBOLIC FUNCTIONS

[Inequalities, differentiation, integration, Maclaurin series, limits, roots of equations, conformal transformation]

9.1 Verify the inequalities

$$\text{sech } x < \text{cosech } x < \text{coth } x, \qquad (x > 0).$$

$$\text{sech } x = \frac{1}{\cosh x} = \frac{2}{e^x + e^{-x}},$$

$$\text{cosech } x = \frac{1}{\sinh x} = \frac{2}{e^x - e^{-x}},$$

$$\coth x = \frac{1}{\tanh x} = \frac{e^x + e^{-x}}{e^x - e^{-x}}.$$

$\operatorname{sech} x < \operatorname{cosech} x$ is equivalent to $\dfrac{2}{e^x + e^{-x}} < \dfrac{2}{e^x - e^{-x}}$

or $4e^{-x} > 0$,   which is so.

$\coth x > \operatorname{cosech} x$ gives $\dfrac{e^x + e^{-x}}{e^x - e^{-x}} > \dfrac{2}{e^x - e^{-x}}$ or $e^x + e^{-x} > 2$.

This inequality may be written as

$$e^{2x} - 2e^x + 1 = (e^x - 1)^2 > 0, \quad \text{which is so.}$$

9.2   Differentiate the functions

(a) $\tan^{-1}(\sinh x)$,   (b) $\log_e(\tanh x)$,   (c) $(\cosh x)^{\sin x}$.

(a) Let $\sinh x = u$. Then

$$\frac{d}{dx}[\tan^{-1}(\sinh x)] = \frac{d}{du}(\tan^{-1} u)\frac{du}{dx}$$

$$= \frac{1}{1+u^2}\cosh x = \frac{\cosh x}{1+\sinh^2 x}$$

$$= \frac{1}{\cosh x} = \operatorname{sech} x \quad (\text{since } \cosh^2 x = 1 + \sinh^2 x).$$

(b) Let $\tanh x = u$. Then

$$\frac{d}{dx}\log_e \tanh x = \frac{d}{du}(\log_e u)\frac{du}{dx}$$

$$= \frac{1}{u}\operatorname{sech}^2 x = \frac{\operatorname{sech}^2 x}{\tanh x} = 2\operatorname{cosech} 2x.$$

(c) Put $f(x) = (\cosh x)^{\sin x}$. Taking logs

$$\log_e f(x) = \sin x \log_e \cosh x,$$

$$\frac{1}{f}\frac{df}{dx} = \cos x \log_e \cosh x + \sin x \tanh x.$$

Therefore

$$\frac{df}{dx} = (\cosh x)^{\sin x}[\cos x \log_e \cosh x + \sin x \tanh x].$$

9.3 Evaluate the integrals

(a) $\int \operatorname{sech} x\, dx,$ (b) $\int_0^{\pi/4} \cosh x \cos x\, dx.$

(a) $\int \operatorname{sech} x\, dx = \int \dfrac{\cosh x}{\cosh^2 x}\, dx = \int \dfrac{d(\sinh x)}{1+\sinh^2 x}$

$$= \int \dfrac{du}{1+u^2} = \tan^{-1} u + C,$$

where $u = \sinh x$ and C is an arbitrary constant of integration.

(b) Let $I = \int_0^{\pi/4} \cosh x \cos x\, dx.$ Integrating by parts we have

$$I = (\cosh x \sin x)_0^{\pi/4} - \int_0^{\pi/4} \sin x \sinh x\, dx.$$

Now

$$\int_0^{\pi/4} \sin x \sinh x\, dx = (-\cos x \sinh x)_0^{\pi/4} + \int_0^{\pi/4} \cos x \cosh x\, dx$$

hence

$$I = (\cosh x \sin x)_0^{\pi/4} - \left[(-\cos x \sinh x)_0^{\pi/4} + \int_0^{\pi/4} \cosh x \cos x\, dx\right]$$

and therefore $2I = 1/\sqrt{2}\cosh(\pi/4) + 1/\sqrt{2}\sinh(\pi/4)$ or $I = (1/2\sqrt{2})e^{\pi/4}.$

9.4 If $y = e^{a \sinh^{-1} x},$ where a is a constant, show that

$$(1+x^2)y'' + xy' - a^2 y = 0.$$

Hence by differentiating this relation n times by Leibnitz's theorem, deduce that

$$(1+x^2)y^{(n+2)} + x(2n+1)y^{(n+1)} + (n^2 - a^2)y = 0.$$

Let $u = \sinh^{-1} x$ so $x = \sinh u.$ Then $dx/du = \cosh u,$ and $y = e^{au}.$ Hence

$$\frac{dy}{dx} = \frac{dy}{du}\frac{du}{dx} = \frac{ae^{au}}{\cosh u} = \frac{ay}{\sqrt{x^2+1}}.$$

Alternatively use the basic result $\sinh^{-1} x = \log_e(x + \sqrt{x^2+1}).$ Hence, by taking logs. first,

$$\frac{1}{y}\frac{dy}{dx} = a\frac{d}{dx}\log_e(x + \sqrt{x^2+1}) = \frac{a}{\sqrt{x^2+1}}.$$

Therefore again

$$\frac{dy}{dx} = \frac{ay}{\sqrt{x^2+1}}. \quad \text{Now} \quad \frac{d^2y}{dx^2} = a\left\{\frac{\sqrt{x^2+1}\,\dfrac{dy}{dx} - \dfrac{x}{\sqrt{x^2+1}}\,y}{x^2+1}\right\}$$

$$= a\left\{\frac{\dfrac{dy}{dx}}{\sqrt{x^2+1}} - \frac{xy}{(x^2+1)^{3/2}}\right\} = a\left\{\frac{ay}{x^2+1} - \frac{xy}{(x^2+1)^{3/2}}\right\}.$$

Hence

$$(1+x^2)y'' = a^2 y - x\frac{ay}{\sqrt{x^2+1}} = a^2 y - x\frac{dy}{dx},$$

which is the required result.

9.5 Show that $x \leqslant \sinh x \leqslant 2x \sinh\left(\frac{1}{2}\right)$ over the range $0 \leqslant x \leqslant \frac{1}{2}$.

Basic result

Maclaurin expansion of $\sinh x = x + \dfrac{x^3}{3!} + \dfrac{x^5}{5!} + \cdots$. Hence $\sinh x \geqslant x$, equality occurring at $x = 0$. To show that $\sinh x \leqslant 2x \sinh\left(\frac{1}{2}\right)$ we write the inequality as $[\sinh\left(\frac{1}{2}\right)]/\left(\frac{1}{2}\right) \geqslant (\sinh x)/x$. But from the series expansion, $(\sinh x)/x$ is easily seen to be a *monotonic increasing* function whose smallest value in the range is unity at $x = 0$. Hence the inequality is satisfied.

9.6 Evaluate

(a) $\displaystyle\lim_{x \to 0}\left\{\frac{\cosh x - 1}{x^2}\right\}$, (b) $\displaystyle\lim_{x \to \infty}(\cosh^{-1} x - \sinh^{-1} x)$.

(a) Maclaurin series for $\cosh x = 1 + \dfrac{x^2}{2!} + \dfrac{x^4}{4!} + \cdots$. Hence

$$\lim_{x \to 0}\left\{\frac{1 + \dfrac{x^2}{2!} + \dfrac{x^4}{4!} + \cdots - 1}{x^2}\right\} = \frac{1}{2}.$$

Basic results

$$\sinh^{-1} x = \log_e (x + \sqrt{x^2 + 1}), \quad \cosh^{-1} x = \log_e (x \pm \sqrt{x^2 - 1}).$$

Hence

$$\lim_{x \to \infty} (\cosh^{-1} x - \sinh^{-1} x) = \lim_{x \to \infty} \left\{ \log_e \left(\frac{x + \sqrt{x^2 - 1}}{x + \sqrt{x^2 + 1}} \right) \right\} = \log_e 1 = 0,$$

taking the positive value of $\cosh^{-1} x$.

9.7   Given the function $y = (1/x) \tanh (x^3/3)$ in $0 < x < \infty$, state its approximate form for small and large x. Show that $y < x^2/3$ and $y < 1/x$. Deduce that $0 < y < 1/3^{1/3}$. Sketch the graph of y against x and verify that it has a stationary point at $(3z/2)^{1/3}$, where $\sinh z = 3z$.

(L.U.)

$\tanh \theta = \dfrac{e^\theta - e^{-\theta}}{e^\theta + e^{-\theta}}$. From this definition we see that $\tanh \theta \to 1$ for large θ, and $\tanh \theta \approx \theta$ for small θ (using the series expansion for e^θ). Also $\tanh \theta < \theta$, $\theta > 0$, and $\tanh \theta < 1$, $\theta > 0$. Hence

$$y = \frac{1}{x} \tanh \frac{x^3}{3} < \frac{1}{x} \frac{x^3}{3} = \frac{x^2}{3} \qquad \text{for } x > 0,$$

and $y < 1/x$ for $x > 0$. The function y lies totally under the curves as shown in Fig. 15. These curves intersect when $1/x = \frac{1}{3}x^2$ or $x = 3^{1/3}$.

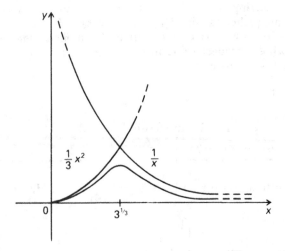

Fig. 15

Hence $0<y<1/3^{1/3}$.

$$y' = -\frac{1}{x^2} \tanh \frac{x^3}{3} + x \, \text{sech}^2 \frac{x^3}{3} = 0 \text{ gives}$$

$$\sinh \frac{x^3}{3} \cosh \frac{x^3}{3} = x^3 \quad \text{or} \quad \sinh \frac{2x^3}{3} = 2x^3$$

(using the double angle formula).

Putting $2x^3/3 = z$ we obtain $\sinh z = 3z$ as required.

9.8 Solve the equations

(a) $\cosh x - \frac{1}{3} \sinh x = 1$,

(b) $\tanh z - \coth z = 2i$, where $z = x + iy$.

(a) $\dfrac{e^x + e^{-x}}{2} - \dfrac{1}{3} \dfrac{e^x - e^{-x}}{2} = 1$, whence $e^{2x} - 3e^x + 2 = 0$. Solving the quadratic for e^x gives $e^x = 1, 2$, or $x = 0, \log_e 2$.

(b) $\left(\dfrac{e^z - e^{-z}}{e^z + e^{-z}}\right) - \left(\dfrac{e^z + e^{-z}}{e^z - e^{-z}}\right) = 2i$,

which gives, after simplification, $e^{4z} - 2ie^{2z} - 1 = 0$, or $(e^{2z} - i)^2 = 0$. Hence $e^{2z} - i = 0$, or

$$2z = \log_e i = \log_e |i| + i \arg i + 2n\pi i, \qquad n = 0, 1, 2, \ldots$$

$$= i\left(\frac{\pi}{2} + 2n\pi\right), \qquad \text{since } |i| = 1.$$

Hence

$$z = i\left(\frac{\pi}{4} + n\pi\right).$$

9.9 Show that the transformation $\omega = \cosh z$ transforms the region defined by $x = 0$, $x = 1$, $y = 0$, $y = \pi/2$ in the z-plane into a quadrant of an ellipse in the ω-plane, and show that the transformation is not conformal at $z = 0$.

$$\omega = u + iv = \cosh z = \cosh(x + iy)$$

$$= \cosh x \cosh iy + \sinh x \sinh iy.$$

Basic results

$$\cosh iy = \frac{e^{iy} + e^{-iy}}{2} = \cos y,$$

$$\sinh iy = \frac{e^{iy} - e^{-iy}}{2} = i \sin y.$$

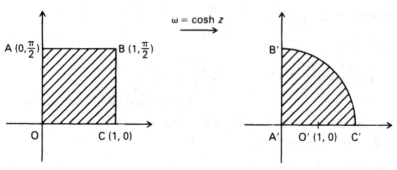

Fig. 16

Hence

$$u + iv = \cosh x \cos y + i \sinh x \sin y,$$

and

$$u = \cosh x \cos y, \qquad v = \sinh x \sin y.$$

The boundary $x = 1$ (CB – see Fig. 16) requires $u = \cosh 1 \cos y$, $v = \sinh 1 \sin y$, which is part of the ellipse $\dfrac{u^2}{\cosh^2 1} + \dfrac{v^2}{\sinh^2 1} = 1$ (C'B' in the ω-plane). Likewise AB transforms into A'B', and so on. The point O in the z-plane becomes O' in the ω-plane. Angles are not preserved at this point since $\omega' = \sinh z = 0$ at $z = 0$, and the transformation is not conformal.

10. PARTIAL DIFFERENTIATION

[Limits, continuity, first and second order partial derivatives, total differentials, chain rules, functions of functions, simple partial differential equations]

10.1 Given

$$f(x, y) = \begin{cases} \dfrac{xy}{x^2 + y^2}, & x \neq 0, y \neq 0, \\ 0, & x = 0, y = 0, \end{cases}$$

evaluate $\lim_{x\to 0} f(x, y)$, $\lim_{y\to 0} f(x, y)$, and obtain the limit as $(x, y) \to (0, 0)$ along the line $y = x$. Hence show that $f(x, y)$ is not continuous at $(0, 0)$.

$$\lim_{x\to 0} \frac{xy}{x^2+y^2} = \lim_{x\to 0} x \lim_{x\to 0} \frac{y}{x^2+y^2} = 0.$$

Likewise $\lim_{y\to 0} xy/(x^2+y^2) = 0$. Along $y = x$,

$$\lim_{(x,y)\to(0,0)} \left(\frac{xy}{x^2+y^2}\right) = \lim_{(x,y)\to(0,0)} \left(\frac{y^2}{2y^2}\right) = \tfrac{1}{2}.$$

Hence $f(0, 0) \neq \lim_{(x,y)\to(0,0)} f(x, y)$, and therefore $f(x, y)$ is not continuous at $(0, 0)$.

10.2 Show that

$$\lim_{x\to 0} \lim_{y\to 0} \frac{x^2-y^2}{x^2+y^2} \neq \lim_{y\to 0} \lim_{x\to 0} \frac{x^2-y^2}{x^2+y^2}.$$

$$\lim_{x\to 0} \lim_{y\to 0} \frac{x^2-y^2}{x^2+y^2} = 1, \qquad \lim_{y\to 0} \lim_{x\to 0} \frac{x^2-y^2}{x^2+y^2} = -1.$$

10.3 Show that $\partial^2 u/\partial x\,\partial y = \partial^2 u/\partial y\,\partial x$ for the following functions:

(a) $u = x^3 + 3xy - y^2$, (b) $u = \log_e (x^2+y)$, (c) $u = x \cos (x/y)$.

(a) $\dfrac{\partial u}{\partial x} = 3x^2+3y$, $\dfrac{\partial^2 u}{\partial y\,\partial x} = 3$, $\dfrac{\partial u}{\partial y} = 3x - 2y$, $\dfrac{\partial^2 u}{\partial x\,\partial y} = 3.$

(b) $\dfrac{\partial u}{\partial x} = \dfrac{2x}{x^2+y}$, $\dfrac{\partial^2 u}{\partial y\,\partial x} = -\dfrac{2x}{(x^2+y)^2}$, $\dfrac{\partial u}{\partial y} = \dfrac{1}{x^2+y}$, $\dfrac{\partial^2 u}{\partial x\,\partial y} = -\dfrac{2x}{(x^2+y)^2}.$

(c) $\dfrac{\partial u}{\partial x} = \cos\left(\dfrac{x}{y}\right) + \dfrac{x}{y}\left(-\sin\left(\dfrac{x}{y}\right)\right)$, $\dfrac{\partial u}{\partial y} = \dfrac{x^2}{y^2}\sin\left(\dfrac{x}{y}\right)$

$\dfrac{\partial^2 u}{\partial y\,\partial x} = \dfrac{x}{y^2}\sin\left(\dfrac{x}{y}\right) + \dfrac{x}{y^2}\sin\left(\dfrac{x}{y}\right) + \dfrac{x^2}{y^3}\cos\left(\dfrac{x}{y}\right) = \dfrac{\partial^2 u}{\partial x\,\partial y}.$

10.4 Given that $u = x \log_e (x^2+y^2) - 2y \tan^{-1} (y/x)$, verify that

$$x\frac{\partial u}{\partial x} + y\frac{\partial u}{\partial y} = u + 2x.$$

$$\frac{\partial u}{\partial x} = x\frac{2x}{x^2+y^2} + \log_e(x^2+y^2) - 2y\frac{1}{1+\left(\frac{y}{x}\right)^2}\left(-\frac{y}{x^2}\right),$$

$$x\frac{\partial u}{\partial x} = \frac{2x^3}{x^2+y^2} + x\log_e(x^2+y^2) + \frac{2xy^2}{x^2+y^2} = 2x + x\log_e(x^2+y^2).$$

$$\frac{\partial u}{\partial y} = \frac{2xy}{x^2+y^2} - 2\tan^{-1}\left(\frac{y}{x}\right) - 2y\frac{1}{1+\left(\frac{y}{x}\right)^2}\left(\frac{1}{x}\right),$$

$$y\frac{\partial u}{\partial y} = \frac{2xy^2}{x^2+y^2} - 2y\tan^{-1}\left(\frac{y}{x}\right) - \frac{2xy^2}{x^2+y^2} = -2y\tan^{-1}\left(\frac{y}{x}\right).$$

Hence

$$x\frac{\partial u}{\partial x} + y\frac{\partial u}{\partial y} = 2x + x\log_e(x^2+y^2) - 2y\tan^{-1}\left(\frac{y}{x}\right) = 2x + u.$$

10.5 Find du/dt (i) by direct substitution of $x(t)$, $y(t)$, and (ii) by using partial derivatives, when

(a) $u = \sin\left(\frac{y}{x}\right)$, $x = t$, $y = t^2$,

(b) $u = x^2y$, $x = \sin t$, $y = t^3$.

(a) $u = \sin\left(\frac{t^2}{t}\right) = \sin t$, $\dfrac{du}{dt} = \cos t$.

Or

$$\frac{du}{dt} = \frac{\partial u}{\partial x}\frac{dx}{dt} + \frac{\partial u}{\partial y}\frac{dy}{dt} = \cos\left(\frac{y}{x}\right)\left(-\frac{y}{x^2}\right)1 + \frac{1}{x}\cos\left(\frac{y}{x}\right)2t$$

$$= \cos t\left[-\frac{y}{x^2} + \frac{2t}{x}\right] = \cos t\left[-\frac{t^2}{t^2} + \frac{2t}{t}\right] = \cos t.$$

(b) $u = x^2y = t^3\sin^2 t$, $\dfrac{du}{dt} = 3t^2\sin^2 t + t^3\sin 2t$.

Or

$$\frac{du}{dt} = \frac{\partial u}{\partial x}\frac{dx}{dt} + \frac{\partial u}{\partial y}\frac{dy}{dt} = 2xy\frac{dx}{dt} + x^2\frac{dy}{dt} = 2xy\cos t + 3x^2t^2$$

$$= 3t^2\sin^2 t + t^3\sin 2t.$$

10.6 Show that if $u = f(x^3 + y^3)$, then $x^2(\partial u/\partial y) = y^2(\partial u/\partial x)$, where f is an arbitrary function.

Let $x^3 + y^3 = v$, say. Then

$$\frac{\partial u}{\partial x} = \frac{df}{dv}\frac{\partial v}{\partial x}, \qquad \frac{\partial u}{\partial y} = \frac{df}{dv}\frac{\partial v}{\partial y}.$$

Hence

$$\frac{\partial u}{\partial x} = f'(v)3x^2, \qquad \frac{\partial u}{\partial y} = f'(v)3y^2.$$

Therefore

$$x^2\frac{\partial u}{\partial y} = 3x^2y^2f'(v), \qquad y^2\frac{\partial u}{\partial x} = 3x^2y^2f'(v),$$

and the result follows.

10.7   Given that $V = f(x - ct) + g(x + ct)$, where f and g are arbitrary functions and c is a constant, prove that

$$\frac{\partial^2 V}{\partial x^2} - \frac{1}{c^2}\frac{\partial^2 V}{\partial t^2} = 0.$$

$$V = f(u) + g(v), \qquad u = x - ct, \qquad v = x + ct.$$

$$\frac{\partial u}{\partial x} = 1, \quad \frac{\partial u}{\partial t} = -c, \quad \frac{\partial v}{\partial x} = 1, \quad \frac{\partial v}{\partial t} = -c.$$

$$\frac{\partial V}{\partial x} = \frac{df}{du}\frac{\partial u}{\partial x} + \frac{df}{dv}\frac{\partial v}{\partial x} = f'(u) + g'(v) \qquad \frac{\partial V}{\partial t} = f'(u)c + g'(v)(-c).$$

Similarly

$$\frac{\partial^2 V}{\partial x^2} = f''(u) + g''(v), \qquad \frac{\partial^2 V}{\partial t^2} = f''(u)c^2 + g''(v)(-c)^2 = c^2\frac{\partial^2 V}{\partial x^2}.$$

10.8   From the implicit relation $f(x, y, z) = 0$, show that

$$\left(\frac{\partial x}{\partial y}\right)_z\left(\frac{\partial y}{\partial z}\right)_x\left(\frac{\partial z}{\partial x}\right)_y = -1.$$

Since $f(x, y, z) = 0$, the total differential

$$df = \frac{\partial f}{\partial x}dx + \frac{\partial f}{\partial y}dy + \frac{\partial f}{\partial z}dz = 0.$$

Therefore when $z = $ constant, $dz = 0$. Hence

$$\left(\frac{\partial x}{\partial y}\right)_z = -\frac{(\partial f/\partial y)}{(\partial f/\partial x)}.$$

Likewise

$$\left(\frac{\partial y}{\partial z}\right)_x = -\frac{(\partial f/\partial z)}{(\partial f/\partial y)} \quad \text{and} \quad \left(\frac{\partial z}{\partial x}\right)_y = -\frac{(\partial f/\partial x)}{(\partial f/\partial z)}.$$

Hence the result.

10.9 Given $x = u + v$, $y = u - v$, show that $\left(\frac{\partial y}{\partial x}\right)_u = -\left(\frac{\partial y}{\partial x}\right)_v$.

$x + y = 2u$. Hence for constant u, $(\partial y/\partial x)_u = -1$.
$x - y = 2v$. Hence for constant v, $(\partial y/\partial x)_v = 1$. Hence the result.

10.10 Show that if $u = u(x, y)$, $x = r \cos \theta$, $y = r \sin \theta$, then

(a) $\left(\frac{\partial u}{\partial x}\right)^2 + \left(\frac{\partial u}{\partial y}\right)^2 = \left(\frac{\partial u}{\partial r}\right)^2 + \frac{1}{r^2}\left(\frac{\partial u}{\partial \theta}\right)^2$,

(b) $x\frac{\partial u}{\partial x} + y\frac{\partial u}{\partial y} = r\frac{\partial u}{\partial r}$,

(c) $-y\frac{\partial u}{\partial x} + x\frac{\partial u}{\partial y} = \frac{\partial u}{\partial \theta}$,

(d) $\frac{\partial^2 u}{\partial x^2} + \frac{\partial^2 u}{\partial y^2} = \frac{\partial^2 u}{\partial r^2} + \frac{1}{r}\frac{\partial u}{\partial r} + \frac{1}{r^2}\frac{\partial^2 u}{\partial \theta^2}$.

(a) $\dfrac{\partial u}{\partial r} = \dfrac{\partial u}{\partial x}\dfrac{\partial x}{\partial r} + \dfrac{\partial u}{\partial y}\dfrac{\partial y}{\partial r} = \cos\theta\dfrac{\partial u}{\partial x} + \sin\theta\dfrac{\partial u}{\partial y}$

$\dfrac{\partial u}{\partial \theta} = \dfrac{\partial u}{\partial x}\dfrac{\partial x}{\partial \theta} + \dfrac{\partial u}{\partial y}\dfrac{\partial y}{\partial \theta}, \quad \dfrac{1}{r}\dfrac{\partial u}{\partial \theta} = -\sin\theta\dfrac{\partial u}{\partial x} + \cos\theta\dfrac{\partial u}{\partial y}$.

Therefore

$$\frac{\partial u}{\partial x} = \cos\theta\frac{\partial u}{\partial r} - \frac{\sin\theta}{r}\frac{\partial u}{\partial \theta}, \quad \frac{\partial u}{\partial y} = \sin\theta\frac{\partial u}{\partial r} + \frac{\cos\theta}{r}\frac{\partial u}{\partial \theta}.$$

Hence

$$\left(\frac{\partial u}{\partial x}\right)^2 + \left(\frac{\partial u}{\partial y}\right)^2 = \left(\frac{\partial u}{\partial r}\right)^2 + \frac{1}{r^2}\left(\frac{\partial u}{\partial \theta}\right)^2.$$

Using the results in (a) for $\dfrac{\partial u}{\partial x}$ and $\dfrac{\partial u}{\partial y}$ the results for (b) and (c) follow.

(d) From (a),

$$\frac{\partial}{\partial x} \equiv \cos\theta\frac{\partial}{\partial r} - \frac{\sin\theta}{r}\frac{\partial}{\partial \theta} \quad \text{and} \quad \frac{\partial}{\partial y} \equiv \sin\theta\frac{\partial}{\partial r} + \frac{\cos\theta}{r}\frac{\partial}{\partial \theta}.$$

Hence

$$\frac{\partial^2 u}{\partial x^2} = \frac{\partial}{\partial x}\left(\frac{\partial u}{\partial x}\right) = \cos\theta\frac{\partial}{\partial r}\left(\cos\theta\frac{\partial u}{\partial r} - \frac{\sin\theta}{r}\frac{\partial u}{\partial\theta}\right)$$

$$-\frac{\sin\theta}{r}\frac{\partial}{\partial\theta}\left(\cos\theta\frac{\partial u}{\partial r} - \frac{\sin\theta}{r}\frac{\partial u}{\partial\theta}\right)$$

$$= \cos^2\theta\frac{\partial^2 u}{\partial r^2} - \frac{2\sin\theta\cos\theta}{r}\frac{\partial^2 u}{\partial r\,\partial\theta} + \frac{\sin^2\theta}{r^2}\frac{\partial^2 u}{\partial\theta^2}$$

$$+\frac{\sin^2\theta}{r}\frac{\partial u}{\partial r} + \frac{2\sin\theta\cos\theta}{r^2}\frac{\partial u}{\partial\theta}.$$

Likewise

$$\frac{\partial^2 u}{\partial y^2} = \frac{\partial}{\partial y}\left(\frac{\partial u}{\partial y}\right) = \sin^2\theta\frac{\partial^2 u}{\partial r^2} + \frac{2\sin\theta\cos\theta}{r}\frac{\partial^2 u}{\partial r\,\partial\theta} + \frac{\cos^2\theta}{r^2}\frac{\partial^2 u}{\partial\theta^2}$$

$$+\frac{\cos^2\theta}{r}\frac{\partial u}{\partial r} - \frac{2\sin\theta\cos\theta}{r^2}\frac{\partial u}{\partial\theta}.$$

Hence finally

$$\frac{\partial^2 u}{\partial x^2} + \frac{\partial^2 u}{\partial y^2} = \frac{\partial^2 u}{\partial r^2} + \frac{1}{r}\frac{\partial u}{\partial r} + \frac{1}{r^2}\frac{\partial^2 u}{\partial\theta^2}.$$

10.11 A function $f(u, v)$ has its variables connected by the relation $v = du/dx$, where u is a function of the independent variable x. Prove that the equation

$$f - v\frac{\partial f}{\partial v} = \text{constant}$$

implies

$$\frac{\partial f}{\partial u} - \frac{d}{dx}\left(\frac{\partial f}{\partial v}\right) = 0,$$

and verify this equivalence for the function $f(u, v) = u\sqrt{(1 + v^2)}$.

(L.U.)

Consider

$$\frac{d}{dx}\left(f - v\frac{\partial f}{\partial v}\right) = \frac{\partial f}{\partial u}\frac{du}{dx} + \frac{\partial f}{\partial v}\frac{dv}{dx} - \frac{dv}{dx}\frac{\partial f}{\partial v} - v\frac{d}{dx}\left(\frac{\partial f}{\partial v}\right)$$

$$= v\left\{\frac{\partial f}{\partial u} - \frac{d}{dx}\left(\frac{\partial f}{\partial v}\right)\right\}.$$

Hence if $f - v\dfrac{\partial f}{\partial v} = \text{constant}$, the result follows.

If $f = u\sqrt{1+v^2}$, $\dfrac{\partial f}{\partial u} = \sqrt{1+v^2}$ and $\dfrac{\partial f}{\partial v} = \dfrac{uv}{\sqrt{1+v^2}}$.

Hence $f - v\dfrac{\partial f}{\partial v} = u\sqrt{1+v^2} - \dfrac{uv^2}{\sqrt{1+v^2}} = \dfrac{u}{\sqrt{1+v^2}} = \text{constant}$.

Now

$$\frac{\partial f}{\partial u} - \frac{d}{dx}\left(\frac{\partial f}{\partial v}\right) = \sqrt{1+v^2} - \frac{d}{dx}\left(\frac{uv}{\sqrt{1+v^2}}\right)$$

$$= \sqrt{1+v^2} - \frac{u}{\sqrt{1+v^2}}\frac{dv}{dx} - \frac{v^2}{\sqrt{1+v^2}}$$

$$+ \frac{uv^2(dv/dx)}{(1+v^2)^{3/2}} \quad \left(\text{using } \frac{du}{dx} = v\right)$$

$$= \frac{1}{\sqrt{1+v^2}} - \frac{u(dv/dx)}{(1+v^2)^{3/2}}.$$

But $\dfrac{u}{\sqrt{1+v^2}} = \text{constant}$, so

$$\frac{d}{dx}\left(\frac{u}{\sqrt{1+v^2}}\right) = 0 = v\left(\frac{1}{\sqrt{1+v^2}} - \frac{u(dv/dx)}{(1+v^2)^{3/2}}\right).$$

Hence $\dfrac{\partial f}{\partial u} - \dfrac{d}{dx}\left(\dfrac{\partial f}{\partial v}\right) = 0$.

10.12   Given the relations

$$k\frac{\partial^2 W}{\partial x^2} = \frac{\partial W}{\partial t}, \qquad u = \frac{x}{2\sqrt{kt}},$$

and k constant, show that, if W is a function of u,

$$\frac{d^2 W}{du^2} + 2u\frac{dW}{du} = 0.$$

Show that a solution of this equation subject to $W = 0$ and $\dfrac{dW}{du} = 1$ when $u = 0$ is given by

$$W = \int_0^u e^{-y^2}\, dy.$$

(L.U.)

If $W' = \dfrac{dW}{du}$, $W'' = \dfrac{d^2 W}{du^2}$ then

$$\frac{\partial W}{\partial x} = \frac{dW}{du}\frac{\partial u}{\partial x} = W' \frac{1}{2\sqrt{kt}},$$

$$\frac{\partial^2 W}{\partial x^2} = W'' \frac{1}{4kt}$$

$$\frac{\partial W}{\partial t} = W' \frac{\partial u}{\partial t} = -\frac{x}{4t^{3/2}\sqrt{k}} W'.$$

Hence $k(\partial^2 W/\partial x^2) = \partial W/\partial t$ implies $W''/4t = -xW'/(4t^{3/2}\sqrt{k})$ or $W'' + 2uW' = 0$. Integrating gives $W' = Ae^{-u^2}$, and the conditions $W' = 1$, $u = 0$ give $A = 1$. Hence

$$W = \int_c^u e^{-y^2}\, dy \qquad (c \text{ is some constant}).$$

Now $W = 0$ when $u = 0$. Hence $c = 0$, and finally therefore

$$W = \int_0^u e^{-y^2}\, dy.$$

11. STATIONARY VALUES OF FUNCTIONS

OF TWO VARIABLES

[Maximum, minimum, and saddle points, constraints, Lagrange multipliers]

11.1 Find the stationary points of the function
$$f(x, y) = x^4 + y^4 - 4a^2 xy,$$
where a is a constant, and determine their nature.

$$\frac{\partial f}{\partial x} = 4x^3 - 4a^2 y = 0, \qquad \frac{\partial f}{\partial y} = 4y^3 - 4a^2 x = 0.$$

Hence $y = x^3/a^2$ and $xa^2 - x^9/a^6 = 0$. Therefore $x(a^8 - x^8) = 0$, giving the roots $x = 0$, a, $-a$. The corresponding values of y are 0, a, $-a$. The stationary points are therefore $(0, 0)$, (a, a), $(-a, -a)$.

Now consider $AB - H^2 = f_{xx}f_{yy} - f_{xy}^2$, where $A = f_{xx} = \partial^2 f/\partial x^2$, $B = f_{yy} = \partial^2 f/\partial y^2$, and $H = f_{xy} = \partial^2 f/(\partial x\, \partial y)$. Evaluating the second partial derivatives at the stationary points, we find at $(0, 0)$, $AB - H^2 = -16a^4 < 0$; hence $(0, 0)$ is a saddle point. At (a, a) and $(-a, -a)$, $AB - H^2 = 128a^4 > 0$ and $A = 12a^2 > 0$. Hence each of these points corresponds to a minimum.

11.2 Find the stationary points of the function

$$f(x, y) = (x^2 + y^2)^2 - 2(x^2 - y^2),$$

and determine their nature. Evaluate the extreme values of the function.

$$\frac{\partial f}{\partial x} = 4x(x^2 + y^2) - 4x = 0, \qquad \frac{\partial f}{\partial y} = 4y(x^2 + y^2) + 4y = 0.$$

Hence $x(x^2 + y^2) - x = 0$, $y(x^2 + y^2) + y = 0$ giving the stationary points $x = 0$, $y = 0$, and $x = \pm 1$, $y = 0$. Evaluating the second partial derivatives A, B and H as in 11.1, we find at $(0, 0)$ $A = -4$, $B = 4$, $H = 0$ so $AB - H^2 = -16 < 0$. Hence $(0, 0)$ is a saddle point. Likewise at $(\pm 1, 0)$, $A = 8$, $B = 8$, $H = 0$, $AB - H^2 = 64 > 0$, $A > 0$. Hence the points $(\pm 1, 0)$ are minima. At $(0, 0)$, $f(x, y) = 0$, and at $(\pm 1, 0)$, $f(x, y) = -1$.

11.3 Given $u = f(x, y)/g(x)$, write down conditions which must be satisfied at a stationary point (x_1, y_1) of the function u. Given that $f(x, y) = F(x) + G(y)$, show that these conditions are equivalent to

$$G'(y_1) = 0 \quad \text{and} \quad \frac{F'(x_1)}{F(x_1) + c_1} = \frac{g'(x_1)}{g(x_1)}, \quad \text{where } c_1 \text{ is to be specified.}$$

Investigate whether the function

$$\frac{x^3 + y^3 - 3y}{2x + 3}$$

has a stationary point in the region $x \geq 0$, $y \geq -1$. If so, how many are there in this region?   (L.U.)

$ug(x) = f(x, y)$. Hence

$$\frac{\partial u}{\partial x} g(x) + ug'(x) = \frac{\partial f}{\partial x} \quad \text{and} \quad \frac{\partial u}{\partial y} g(x) = \frac{\partial f}{\partial y}.$$

At the stationary point (x_1, y_1), $\partial u/\partial x = 0$, $\partial u/\partial y = 0$ so $ug'(x) = \partial f/\partial x$ and $\partial f/\partial y = 0$. Hence if $f(x, y) = F(x) + G(y)$, then $(\partial f/\partial y)_{x_1, y_1} = 0$

gives $G'(y_1) = 0$, and $(\partial f/\partial x)_{x_1,y_1} = ug'(x_1)$ gives

$$F'(x_1) = \frac{g'(x_1)}{g(x_1)} (F(x_1) + G(y_1)),$$

where $G(y_1)$ may be taken as c_1. Hence the result.

For $u = \dfrac{x^3 + y^3 - 3y}{2x + 3}$, let $f(x, y) = x^3 + y^3 - 3y$, $g(x) = 2x + 3$, and write $F(x) = x^3$, $G(y) = y^3 - 3y$. Then $G'(y_1) = 0$ gives $3y_1^2 - 3 = 0$, $y_1 = \pm 1$. Hence $G(y_1) = c_1 = \mp 2$. Also $F'(x_1) = 3x_1^2 = \dfrac{2}{2x_1 + 3} (x_1^3 \mp 2)$, using the result already proved. When $y_1 = +1$ we have to take the negative sign of c_1, so $3x_1^2 = \dfrac{2}{2x_1 + 3} (x_1^3 - 2)$ or $4x_1^3 + 9x_1^2 - 4 = 0$. By sketching the graph of $4x_1^3 + 9x_1^2 - 4$ this equation can be shown to have one positive root and two negative roots. When $y_1 = -1$ we require the positive sign of c_1 so $4x_1^3 + 9x_1^2 + 4 = 0$. This equation has only one real root which can be shown graphically to be negative. Hence there is just one stationary point in the specified region $x \geq 0$, $y \geq -1$ at $(x_1, 1)$, where x_1 is the positive root of $4x_1^3 + 9x_1^2 - 4 = 0$. (N.B. this is approximately 0.595).

11.4 Find the maximum distance from the origin $(0, 0)$ to the curve $3x^2 + 3y^2 + 4xy - 2 = 0$ using Lagrange multipliers.

The distance l from the origin to the curve is given by $l^2 = x^2 + y^2 = f(x, y)$, say. We wish to find the extreme value of this function subject to the constraint

$$g(x, y) = 3x^2 + 3y^2 + 4xy - 2 = 0.$$

Using the Lagrange multiplier technique, we form $f + \lambda g$, where λ is some parameter, and require

$$\frac{\partial}{\partial x} (f + \lambda g) = 2x + \lambda(6x + 4y) = 0,$$

$$\frac{\partial}{\partial y} (f + \lambda g) = 2y + \lambda(6y + 4x) = 0,$$

which gives

$$x(1 + 3\lambda) + 2\lambda y = 0, \qquad 2\lambda x + y(1 + 3\lambda) = 0.$$

Non-trivial (i.e. non-zero) solutions for x and y exist only if

$$(1 + 3\lambda)^2 = 4\lambda^2,$$

whence $\lambda = -1$ or $\lambda = -\frac{1}{5}$.

Case $\lambda = -1$. The derivative equations (above) are now

$$x - (3x + 2y) = 0, \qquad y - (3y + 2x) = 0,$$

whence $y = x$.
Case $\lambda = -\frac{1}{5}$. Likewise

$$x - \tfrac{1}{5}(3x + 2y) = 0, \qquad y - \tfrac{1}{5}(3y + 2x) = 0,$$

whence $y = -x$.
When $y = x$, $g(x, y) = 0$ gives $10x^2 - 2 = 0$ or $x = \pm 1/\sqrt{5}$.
When $y = -x$, $g(x, y) = 0$ gives $2x^2 - 2 = 0$ or $x = \pm 1$.
Hence stationary points are $(1, -1)$, $(-1, 1)$, $(1/\sqrt{5}, 1/\sqrt{5})$, $(-1/\sqrt{5}, -1/\sqrt{5})$, and $l^2 = 2$ for the first two points, and $l^2 = \frac{2}{5}$ for the second two points. Hence maximum distance is $l = \sqrt{2}$ obtained by the line $y = -x$.

11.5   Find the stationary points of the function $f = x^2 + y^2 + z$ subject to the condition $x^2 - z^2 = 1$.

Method 1. $x^2 = 1 + z^2$. Therefore $f = (1 + z^2) + y^2 + z$. Now $\partial f / \partial y = 2y = 0$, $\partial f / \partial z = 2z + 1 = 0$. Hence $y = 0$, $z = -\frac{1}{2}$, whence $x^2 = 1 + z^2 = 1 + \frac{1}{4} = \frac{5}{4}$, or $x = \pm \sqrt{5}/2$. The stationary points are therefore

$$\left(\frac{\sqrt{5}}{2}, 0, -\frac{1}{2} \right) \quad \text{and} \quad \left(-\frac{\sqrt{5}}{2}, 0, -\frac{1}{2} \right).$$

Method 2.   Lagrange multiplier technique. Let $g = x^2 - z^2 - 1$. Then

$$\frac{\partial}{\partial x}(f + \lambda g) = \frac{\partial}{\partial x}(x^2 + y^2 + z + \lambda(x^2 - z^2 - 1)) = 2x + 2\lambda x = 0,$$

$$\frac{\partial}{\partial y}(f + \lambda g) = \frac{\partial}{\partial y}(x^2 + y^2 + z + \lambda(x^2 - z^2 - 1)) = 2y = 0,$$

$$\frac{\partial}{\partial z}(f + \lambda g) = \frac{\partial}{\partial z}(x^2 + y^2 + z + \lambda(x^2 - z^2 - 1)) = 1 - 2\lambda z = 0.$$

Hence $z = 1/2\lambda$, $y = 0$, $x(1 + \lambda) = 0$. Finally imposing $g = x^2 - z^2 - 1 = 0$ leads to the solutions $\lambda = -1$, $z = -\frac{1}{2}$, $y = 0$, $x = \pm \sqrt{5}/2$, as before.

12. DETERMINANTS AND DIFFERENCE

EQUATIONS

[Evaluation and factorization of determinants, Cramer's rule, functional dependence, Jacobian determinants, linear dependence, Wronskian determinants, difference equations (linear and non-linear)]

12.1 Show that the determinant

$$\begin{vmatrix} x-b & 0 & b-x \\ 2x & x+b & 2x \\ x+b & 2b & x+b \end{vmatrix}$$

is of the form $k(x-a)^3$, and find a and k.

Adding the third column to the first column gives

$$\begin{vmatrix} 0 & 0 & b-x \\ 4x & x+b & 2x \\ 2(x+b) & 2b & x+b \end{vmatrix}.$$

Now expand along the first row. Then $(b-x)[8xb-2(x+b)^2] = 2(x-b)^3$. Hence $a=b$ and $k=2$.

12.2 Factorize

$$\text{(a)} \quad \begin{vmatrix} 1 & 1 & 1 \\ x & y & z \\ x^3 & y^3 & z^3 \end{vmatrix} \qquad \text{(b)} \quad \begin{vmatrix} a^3 & b^3 & c^3 \\ a^2 & b^2 & c^2 \\ b+c & c+a & a+b \end{vmatrix}.$$

(a) Subtracting the second column from the first column gives

$$\begin{vmatrix} 0 & 1 & 1 \\ x-y & y & z \\ x^3-y^3 & y^3 & z^3 \end{vmatrix}.$$

Subtracting the third column from the second column gives

$$\begin{vmatrix} 0 & 0 & 1 \\ x-y & y-z & z \\ x^3-y^3 & y^3-z^3 & z^3 \end{vmatrix}.$$

Hence expanding along the first row gives

$$(x-y)(y^3-z^3)-(y-z)(x^3-y^3)=(x-y)(y-z)(z-x)(x+y+z).$$

(b) If $a = b$, two columns are identical and hence the determinant is zero. Consequently $a-b$ is a factor. Likewise, if $b=c$ or $c=a$ the determinant vanishes. Hence $b-c$ and $c-a$ are also factors. Similarly, if $a=-b$ expansion shows that the determinant again vanishes. Hence $a+b$ is a factor, as are $b+c$ and $c+a$ by the same argument. Hence the determinant may be factorized as

$$k(a-b)(b-c)(c-a)(a+b)(b+c)(c+a),$$

where, by looking at the coefficient of a^4b^2 (say) in the expansion and comparing with the factorized form, $k=-1$.

12.3 Find values of λ for which the equations

$$2\lambda x + 2y + 3z = 0,$$
$$2x + \lambda y + z = 0,$$
$$2\lambda x + y + \lambda z = 0,$$

have non-trivial solutions. Show also that when $\lambda = 1$, these equations have a solution consistent with

$$5x - 3y - z = 10,$$

and find that solution. (L.U.).

Using Cramer's rule,

$$x = \frac{\begin{vmatrix} 0 & 2 & 3 \\ 0 & \lambda & 1 \\ 0 & 1 & \lambda \end{vmatrix}}{\begin{vmatrix} 2\lambda & 2 & 3 \\ 2 & \lambda & 1 \\ 2\lambda & 1 & \lambda \end{vmatrix}} \qquad y = \frac{\begin{vmatrix} 2\lambda & 0 & 3 \\ 2 & 0 & 1 \\ 2\lambda & 0 & \lambda \end{vmatrix}}{\begin{vmatrix} 2\lambda & 2 & 3 \\ 2 & \lambda & 1 \\ 2\lambda & 1 & \lambda \end{vmatrix}} \quad \text{and} \quad z = \frac{\begin{vmatrix} 2\lambda & 2 & 0 \\ 2 & \lambda & 0 \\ 2\lambda & 1 & 0 \end{vmatrix}}{\begin{vmatrix} 2\lambda & 2 & 3 \\ 2 & \lambda & 1 \\ 2\lambda & 1 & \lambda \end{vmatrix}}$$

The determinants in the numerators are all zero, hence the solutions are $x=0$, $y=0$, $z=0$ (i.e. the trivial solutions) *unless* the determinant in the denominator vanishes. Hence non-trivial solutions are possibly only if

$$\begin{vmatrix} 2\lambda & 2 & 3 \\ 2 & \lambda & 1 \\ 2\lambda & 1 & \lambda \end{vmatrix} = 0,$$

which gives three possible values of λ, $\lambda = 3$, ± 1. For $\lambda = 1$ the equations become

$$2\left(\frac{x}{z}\right) + 2\left(\frac{y}{z}\right) = -3, \qquad 2\left(\frac{x}{z}\right) + \left(\frac{y}{z}\right) = -1,$$

whence $x = -\mu$, $y = 4\mu$, $z = -2\mu$, where μ is a parameter. To be consistent with the equation $5x - 3y - z = 10$ we find $\mu = -\frac{2}{3}$. Hence the required solutions are

$$x = \tfrac{2}{3}, \quad y = -\tfrac{8}{3}, \quad z = \tfrac{4}{3}.$$

12.4 Given the transformation $x = u - 2v$, $y = 2u + v$ (a) obtain the inverse transformation for u and v in terms of x and y, (b) evaluate the Jacobians

$$J = \frac{\partial(u, v)}{\partial(x, y)}, \qquad J^{-1} = \frac{\partial(x, y)}{\partial(u, v)},$$

and verify $JJ^{-1} = 1$.

Find the Jacobian $\dfrac{\partial(\xi, \eta)}{\partial(x, y)}$, where $x = c \cosh \xi \cos \eta$, $y = c \sinh \xi \sin \eta$ (c is a constant).

Solving for u and v we find $u = \tfrac{1}{5}(x + 2y)$, $v = \tfrac{1}{5}(y - 2x)$. Now $\dfrac{\partial x}{\partial u} = 1$,

$\dfrac{\partial y}{\partial u} = 2$, $\dfrac{\partial x}{\partial v} = -2$, $\dfrac{\partial y}{\partial v} = 1$, $\dfrac{\partial u}{\partial x} = \tfrac{1}{5}$, $\dfrac{\partial u}{\partial y} = \tfrac{2}{5}$, $\dfrac{\partial v}{\partial x} = -\tfrac{2}{5}$, and $\dfrac{\partial v}{\partial y} = \tfrac{1}{5}$. Hence

$$J = \begin{vmatrix} \dfrac{\partial u}{\partial x} & \dfrac{\partial v}{\partial x} \\ \dfrac{\partial u}{\partial y} & \dfrac{\partial v}{\partial y} \end{vmatrix} = \begin{vmatrix} \tfrac{1}{5} & -\tfrac{2}{5} \\ \tfrac{2}{5} & \tfrac{1}{5} \end{vmatrix} = \tfrac{1}{5},$$

and

$$J^{-1} = \begin{vmatrix} \dfrac{\partial x}{\partial u} & \dfrac{\partial y}{\partial u} \\ \dfrac{\partial x}{\partial v} & \dfrac{\partial y}{\partial v} \end{vmatrix} = \begin{vmatrix} 1 & 2 \\ -2 & 1 \end{vmatrix} = 5.$$

Consequently $JJ^{-1} = 1$.

Similarly

$$\frac{\partial(\xi, \eta)}{\partial(x, y)} = 1 \Big/ \frac{\partial(x, y)}{\partial(\xi, \eta)}.$$

Now

$$\frac{\partial(x, y)}{\partial(\xi, \eta)} = \begin{vmatrix} c \sinh \xi \cos \eta & c \cosh \xi \sin \eta \\ -c \cosh \xi \sin \eta & c \sinh \xi \cos \eta \end{vmatrix} = c^2(\cosh^2 \xi - \cos^2 \eta).$$

Hence

$$\frac{\partial(\xi, \eta)}{\partial(x, y)} = \frac{1}{c^2(\cosh^2 \xi - \cos^2 \eta)}.$$

12.5 If $x = r \cos \theta$, $y = r \sin \theta$, show that $J_1 = \partial(r, \theta)/\partial(x, y) = 1/r$. If $u = e^r \cos \theta$, $v = e^r \sin \theta$, find $J_2 = \partial(u, v)/\partial(r, \theta)$. Hence using the result $J_3 = \partial(u, v)/\partial(x, y) = J_2 J_1$, evaluate J_3.

$$J_1 = \begin{vmatrix} \dfrac{\partial r}{\partial x} & \dfrac{\partial \theta}{\partial x} \\ \dfrac{\partial r}{\partial y} & \dfrac{\partial \theta}{\partial y} \end{vmatrix}. \quad r^2 = x^2 + y^2, \ \tan \theta = \frac{y}{x}.$$

Hence

$$J_1 = \begin{vmatrix} \dfrac{x}{r} & -\dfrac{y}{r^2} \\ \dfrac{y}{r} & \dfrac{x}{r^2} \end{vmatrix} = \frac{1}{r}.$$

$$J_2 = \begin{vmatrix} \dfrac{\partial u}{\partial r} & \dfrac{\partial v}{\partial r} \\ \dfrac{\partial u}{\partial \theta} & \dfrac{\partial v}{\partial \theta} \end{vmatrix} = \begin{vmatrix} e^r \cos \theta & e^r \sin \theta \\ -e^r \sin \theta & e^r \cos \theta \end{vmatrix} = e^r.$$

Therefore $J_3 = \dfrac{e^r}{r}$.

12.6 Determine whether, in the two following cases, u and v are functionally dependent or not:
(a): $u = x \log_e y$, $v = y \log_e x$,
(b) $u = \sinh x \cosh y + \cosh x \sinh y$, $v = (x + y)^2$.

(a) $J = \dfrac{\partial(u, v)}{\partial(x, y)} = \begin{vmatrix} \log_e y & \dfrac{y}{x} \\ \dfrac{x}{y} & \log_e x \end{vmatrix} \neq 0.$

Hence u and v are not functionally dependent

(b) $J = \begin{vmatrix} \cosh(x+y) & \cosh(x+y) \\ 2(x+y) & 2(x+y) \end{vmatrix} = 0.$

Hence functionally dependent. In fact, $u = \sinh(x+y) = \sinh\sqrt{v}$.

12.7 Determine which of the following sets of functions are linearly dependent, and find the relations between the functions in these cases:
(a) $u = x$, $v = x^2$, $w = x^3$,
(b) $u = e^x$, $v = e^{-x}$, $w = 2\cosh x$.

Basic result

A set of functions $f_1(x), f_2(x), \ldots, f_n(x)$ is linearly dependent if $c_1 f_1 + c_2 f_2 + \cdots + c_n f_n = 0$ for all x and for values of the constants c_i which are not all zero. Hence the functions are linearly dependent if their Wronskian determinant vanishes.
(a) The Wronskian W is

$$W = \begin{vmatrix} x & x^2 & x^3 \\ 1 & 2x & 3x^2 \\ 0 & 2 & 6x \end{vmatrix} = 2x^3 \neq 0 \quad \text{(for all } x\text{)}.$$

Hence not linearly dependent.

(b) $W = \begin{vmatrix} e^x & e^{-x} & 2\cosh x \\ e^x & -e^{-x} & 2\sinh x \\ e^x & e^{-x} & 2\cosh x \end{vmatrix} = 0,$

since two rows of the determinant are identical. Hence linearly dependent. Clearly $c_1 e^x + c_2 e^{-x} + c_3 \cosh x = 0$, where $c_1 = 1$, $c_2 = 1$ and $c_3 = -2$.

12.8 Given that D_n is the nth order determinant defined by

$$D_n = \begin{vmatrix} 2\cosh\alpha & 1 & 0 & \cdots & 0 & 0 \\ 1 & 2\cosh\alpha & 1 & \cdots & 0 & 0 \\ 0 & 1 & 2\cosh\alpha & \cdots & 0 & 0 \\ \vdots & \vdots & \vdots & \ddots & \vdots & \vdots \\ 0 & 0 & 0 & \cdots & 2\cosh\alpha & 1 \\ 0 & 0 & 0 & \cdots & 1 & 2\cosh\alpha \end{vmatrix}$$

show that D_n satisfies the *difference* equation

$$D_n = 2 \cos \alpha \, D_{n-1} + D_{n-2} = 0$$

for $n \geqslant 3$. Hence deduce that

$$D_n = \frac{\sinh (n+1)\alpha}{\sinh \alpha}.$$

Expanding down the first column gives $D_n = 2 \cosh \alpha D_{n-1} - D_{n-2}$, where $n \geqslant 3$. Hence the result.

Let $D_n = x^n$. Then from this difference equation $x^2 - 2 \cosh \alpha x + 1 = 0$ we find

$$x = \frac{2 \cosh \alpha \pm \sqrt{4 \cosh^2 \alpha - 4}}{2} = \cosh \alpha \pm \sinh \alpha,$$

whence $x = e^\alpha$ or $e^{-\alpha}$. Hence the general solution is $D_n = Ae^{n\alpha} + Be^{-n\alpha}$, or equivalently

$$D_n = E \cosh n\alpha + F \sinh n\alpha, \quad (E, F \text{ arbitrary constants}).$$

Now

$$D_1 = 2 \cosh \alpha = E \cosh \alpha + F \sinh \alpha$$

and

$$D_2 = 4 \cosh^2 \alpha - 1 = E \cosh 2\alpha + F \sinh 2\alpha.$$

Hence $E = 1$, $F = \coth \alpha$, and finally

$$D_n = 1 \cosh n\alpha + \coth \alpha \sinh n\alpha = \frac{\sinh (n+1)\alpha}{\sinh \alpha}.$$

12.9 Given that $u_{n+2} - 4u_{n+1} + 5u_n = 0$, where $n = 0, 1, 2, \ldots$, solve for u_n given $u_0 = 1$, $u_1 = 2$.

Let $u_n = x^n$. Then $x^2 - 4x + 5 = 0$, whence $x = 2 \pm i$. Hence the two solutions are $x = 2 + i$ and $x = 2 - i$. Writing $2 + i = re^{i\theta}$, we find $r = \sqrt{5}$, $\sin \theta = 1/\sqrt{5}$ and $\cos \theta = 2/\sqrt{5}$. Similarly $2 - i = re^{-i\theta}$. Hence the two solutions are

$$u_n = (\sqrt{5}\, e^{i\theta})^n = 5^{n/2}(\cos n\theta + i \sin n\theta)$$

and

$$u_n = (\sqrt{5}\, e^{-i\theta})^n = 5^{n/2}(\cos n\theta - i \sin n\theta).$$

The general solution is therefore of the form

$$u_n = A \cdot 5^{n/2}(\cos n\theta + i \sin n\theta) + B \cdot 5^{n/2}(\cos n\theta - i \sin n\theta).$$

where A and B are arbitrary constants, or equivalently as

$$u_n = 5^{n/2}(E \cos n\theta + F \sin n\theta),$$

where E and F are arbitrary constants. Given that $u_0 = 1$ we find $E = 1$, and $u_1 = 2$ gives $2 = \sqrt{5}\,(2/\sqrt{5} + F/\sqrt{5}\,)$, whence $F = 0$. Hence the solution satisfying the given conditions is

$$u_n = 5^{n/2} \cos n\theta,$$

where $\cos \theta = 2/\sqrt{5}$.

12.10 Solve the difference equations

$$u_{n+1} - 4u_n - u_{n-1} = 16,$$
$$v_{n+1} - 2v_n + v_{n-1} = 0,$$

where $n = 1, 2, 3, \ldots$, given that

$$u_0 = -4,$$
$$v_0 = 0,$$
$$u_1 + v_1 = 0,$$
$$u_2 + v_2 = 0.$$
(L.U.)

Consider first the homogeneous form of the u equation (i.e. right-hand side zero). Let $u_n = x^n$. Then $x^2 - 4x - 1 = 0$, whence $x = 2 \pm \sqrt{5}$. A particular solution of the inhomogeneous u equation is (by observation) $u_n = -4$. Hence the general solution is of the form

$$u_n = A(2+\sqrt{5})^n + B(2-\sqrt{5})^n - 4.$$

Now $u_0 = -4$ implies that $A + B = 0$ or $A = -B$. The solution of the v equation is $v_n = Cn + D$, where C and D are arbitrary constants. The condition $v_0 = 0$ implies that $D = 0$. The condition $u_1 + v_1 = 0$ gives $2\sqrt{5}A - 4 + C = 0$, and $u_2 + v_2 = 0$ gives $8\sqrt{5}A - 4 + 2C = 0$. Hence $C = 6$, $A = -1/\sqrt{5}$, and the solutions are

$$u_n = \frac{1}{\sqrt{5}}(2-\sqrt{5})^n - \frac{1}{\sqrt{5}}(2+\sqrt{5})^n - 4, \qquad v_n = 6n.$$

12.11 Solve the non-linear difference equation

$$u_n u_{n+2} = u_n^2, \qquad n = 0, 1, 2, 3, \ldots \qquad \text{given } u_0 = 1, \ u_1 = e.$$

Put $z_n = \log_e u_n$. Then taking logs., $z_n + z_{n-2} - 2z_{n+1} = 0$. Now let $z_n = x^n$, whence $x = 1$ (twice). The solution for z_n in the case of a repeated root is $z_n = (C + Dn)1^n$, where C and D are arbitrary constants of integration. Given $u_0 = 1$, $u_1 = e$, we find $z_0 = 0$, $z_1 = 1$. Hence $C = 0$, $D = 1$, and accordingly $z_n = n$. Hence $u_n = e^n$.

12.12   Given that $u_{n+1} = u_n^2 - u_n + 1$, $n = 0, 1, 2, \ldots$ and $u_0 = 2$, show that $\sum_{n=0}^{\infty} 1/u_n = 1$.

Now

$$\frac{1}{u_{n+1} - 1} = \frac{1}{u_n^2 - u_n} = -\frac{1}{u_n} + \frac{1}{u_n - 1},$$

by partial fractions. Therefore

$$\sum_{n=0}^{\infty} \frac{1}{u_n} = \sum_{n=0}^{\infty} \left(\frac{1}{u_n - 1} - \frac{1}{u_{n+1} - 1} \right) = \frac{1}{u_0 - 1} + \lim_{n \to \infty} \left(\frac{1}{u_{n+1} - 1} \right) = 1,$$

since, from the difference equation and initial condition, $u_{n+1} \to \infty$ as $n \to \infty$.

---

13. VECTOR ALGEBRA

[Scalar and vector products, inequalities, straight lines and planes, triple products, vector equations, differentiation of vectors]

13.1   The position vectors of the four points A, B, C, D are **a**, **b**, **a** − 3**b**, and 2**a** + **b**, respectively. Express $\overrightarrow{AB}$, $\overrightarrow{BD}$, $\overrightarrow{DC}$, $\overrightarrow{CA}$ in terms of **a** and **b**, and verify geometrically that the sum of these four vectors is zero.

$\overrightarrow{AB} = \mathbf{b} - \mathbf{a}$, $\overrightarrow{BD} = 2\mathbf{a}$, $\overrightarrow{DC} = -\mathbf{a} - 4\mathbf{b}$, $\overrightarrow{CA} = 3\mathbf{b}$.

The path represented by the sum of the displacements, $\overrightarrow{AB} + \overrightarrow{BD} + \overrightarrow{DC} + \overrightarrow{CA}$, starts at A and finishes there (see Fig. 17). Hence the vector sum is zero.

13.2   Find the scalar and vector products when
(a) $\mathbf{a} = \mathbf{i} - \mathbf{j} + \mathbf{k}$, $\mathbf{b} = 2\mathbf{i} + 3\mathbf{j} - \mathbf{k}$,
(b) $\mathbf{a} = \mathbf{i} + \mathbf{j} + 2\mathbf{k}$, $\mathbf{b} = 3\mathbf{i} - \mathbf{j} - \mathbf{k}$.

For the three orthogonal unit vectors **i**, **j**, **k** we have the basic results

$$\mathbf{i} \cdot \mathbf{i} = \mathbf{j} \cdot \mathbf{j} = \mathbf{k} \cdot \mathbf{k} = 1, \qquad \mathbf{i} \cdot \mathbf{j} = \mathbf{j} \cdot \mathbf{k} = \mathbf{k} \cdot \mathbf{i} = 0, \quad (\mathbf{i} \cdot \mathbf{j} = \mathbf{j} \cdot \mathbf{i}, \text{ etc.})$$

$$\mathbf{i} \wedge \mathbf{i} = \mathbf{j} \wedge \mathbf{j} = \mathbf{k} \wedge \mathbf{k} = 0, \qquad \mathbf{i} \wedge \mathbf{j} = \mathbf{k}, \qquad \mathbf{j} \wedge \mathbf{k} = \mathbf{i}, \qquad \mathbf{k} \wedge \mathbf{i} = \mathbf{j},$$
$$(\mathbf{i} \wedge \mathbf{j} = -\mathbf{j} \wedge \mathbf{i}, \text{ etc.})$$

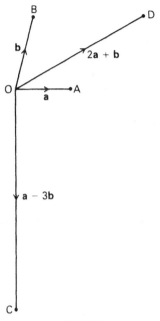

Fig. 17

(a) $\mathbf{a} \cdot \mathbf{b} = (\mathbf{i} - \mathbf{j} + \mathbf{k}) \cdot (2\mathbf{i} + 3\mathbf{j} - \mathbf{k}) = 2 - 3 - 1 = -2.$

$\mathbf{a} \wedge \mathbf{b} = (\mathbf{i} - \mathbf{j} + \mathbf{k}) \wedge (2\mathbf{i} + 3\mathbf{j} - \mathbf{k})$

$\quad = 2(\mathbf{i} \wedge \mathbf{i}) - 2(\mathbf{j} \wedge \mathbf{i}) + 2(\mathbf{k} \wedge \mathbf{i}) + 3(\mathbf{i} \wedge \mathbf{j})$

$\quad\quad - 3(\mathbf{j} \wedge \mathbf{j}) + 3(\mathbf{k} \wedge \mathbf{j}) - (\mathbf{i} \wedge \mathbf{k}) + (\mathbf{j} \wedge \mathbf{k}) - (\mathbf{k} \wedge \mathbf{k})$

$\quad = -2\mathbf{i} + 3\mathbf{j} + 5\mathbf{k}.$

(b) $\mathbf{a} \cdot \mathbf{b} = 0, \qquad \mathbf{a} \wedge \mathbf{b} = \mathbf{i} + 7\mathbf{j} - 4\mathbf{k}.$

13.3 Determine the constant c so that $\mathbf{u} = 3\mathbf{i} - 2\mathbf{j}$ and $\mathbf{v} = 4\mathbf{i} + c\mathbf{j}$ are perpendicular.

Scalar product $\mathbf{u} \cdot \mathbf{v} = |\mathbf{u}|\,|\mathbf{v}| \cos \theta$, where θ is the angle between $\mathbf{u}$ and $\mathbf{v}$. $|\mathbf{u}| \neq 0$, $|\mathbf{v}| \neq 0$. Hence with $\theta = \pi/2$, $\mathbf{u} \cdot \mathbf{v} = 0$, and therefore $(3\mathbf{i} - 2\mathbf{j}) \cdot (4\mathbf{i} + c\mathbf{j}) = 0$, giving $c = 6$.

13.4 Find the lengths of the vectors $2\mathbf{i} - \mathbf{j} + 2\mathbf{k}$, $5\mathbf{i} + 3\mathbf{j} - \mathbf{k}$ and the angle between them.

If $\mathbf{a} = 2\mathbf{i} - \mathbf{j} + 2\mathbf{k}$, then $|\mathbf{a}|^2 = 2^2 + (-1)^2 + 2^2 = 9$, $|\mathbf{a}| = 3$. Likewise, if $\mathbf{b} = 5\mathbf{i} + 3\mathbf{j} - \mathbf{k}$, then $|\mathbf{b}|^2 = 5^2 + 3^2 + (-1)^2 = 35$, $|\mathbf{b}| = \sqrt{35}$. $\mathbf{a} \cdot \mathbf{b} = |\mathbf{a}|\,|\mathbf{b}| \cos \theta$, where θ is the angle between the vectors. Now $\mathbf{a} \cdot \mathbf{b} =$

$2 \times 5 + (-1) \times 3 + 2 \times (-1) = 5$. Hence

$$\cos \theta = \frac{5}{3\sqrt{35}}, \quad \theta = \cos^{-1}\left(\frac{5}{3\sqrt{35}}\right).$$

13.5 Verify the inequalities $\|\mathbf{a}\| - \|\mathbf{b}\|\| \leqslant |\mathbf{a}+\mathbf{b}| \leqslant |\mathbf{a}| + |\mathbf{b}|$ for the vectors $\mathbf{a} = \mathbf{i} - 2\mathbf{j} - \mathbf{k}$, $\mathbf{b} = 2\mathbf{i} - 3\mathbf{j} + \mathbf{k}$.

$|\mathbf{a}|^2 = 6$, $|\mathbf{b}|^2 = 14$, $\mathbf{a} + \mathbf{b} = 3\mathbf{i} - 5\mathbf{j}$, $|\mathbf{a}+\mathbf{b}|^2 = 3^2 + 5^2 = 34$. Hence $|\mathbf{a}| = \sqrt{6}$, $|\mathbf{b}| = \sqrt{14}$, $|\mathbf{a}+\mathbf{b}| = \sqrt{34}$.

$$\|\mathbf{a}| - |\mathbf{b}\|| = |2.45 - 3.74| = 1.29, \qquad |\mathbf{a}+\mathbf{b}| = 5.83, \qquad |\mathbf{a}| + |\mathbf{b}| = 6.19.$$

The results follow.

13.6 The position vectors of the foci, A and B, of an ellipse are respectively $\mathbf{c}$ and $-\mathbf{c}$, and the length of the major axis is $2a$. Prove that the equation of the ellipse can be written in the form

$$a^4 - a^2(r^2 + c^2) + (\mathbf{r} \cdot \mathbf{c})^2 = 0,$$

where $\mathbf{r}$ is the position vector of any point on the ellipse with respect to its centre, and $r = |\mathbf{r}|$, $c = |\mathbf{c}|$.

Use the basic property of an ellipse that the sum of the distances from any point P on the ellipse to the foci is equal to the major axis (see Fig. 18). Hence

$$|\mathbf{r}+\mathbf{c}| + |\mathbf{r}-\mathbf{c}| = 2a,$$
$$|\mathbf{r}+\mathbf{c}| = 2a - |\mathbf{r}-\mathbf{c}|,$$
$$|\mathbf{r}+\mathbf{c}|^2 = 4a^2 - 4a\,|\mathbf{r}-\mathbf{c}| + |\mathbf{r}-\mathbf{c}|^2.$$

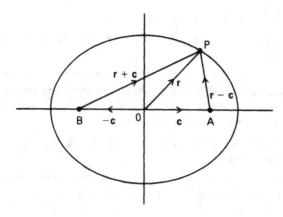

Fig. 18

Therefore

$$(\mathbf{r}+\mathbf{c}) \cdot (\mathbf{r}+\mathbf{c}) = |\mathbf{r}+\mathbf{c}|^2 = 4a^2 - 4a\,|\mathbf{r}-\mathbf{c}| + (\mathbf{r}-\mathbf{c}) \cdot (\mathbf{r}-\mathbf{c}).$$

Hence

$$\mathbf{r} \cdot \mathbf{r} + 2\mathbf{c} \cdot \mathbf{r} + \mathbf{c} \cdot \mathbf{c} = 4a^2 - 4a\,|\mathbf{r}-\mathbf{c}| + \mathbf{r} \cdot \mathbf{r} - 2\mathbf{c} \cdot \mathbf{r} + \mathbf{c} \cdot \mathbf{c}$$

or

$$a^2 - \mathbf{c} \cdot \mathbf{r} = a\,|\mathbf{r}-\mathbf{c}|.$$

Squaring both sides

$$a^4 - 2a^2(\mathbf{c} \cdot \mathbf{r}) + (\mathbf{c} \cdot \mathbf{r})^2 = a^2\,|\mathbf{r}-\mathbf{c}|^2 = a^2(\mathbf{r}-\mathbf{c}) \cdot (\mathbf{r}-\mathbf{c})$$
$$= a^2(r^2 - 2\mathbf{c} \cdot \mathbf{r} + c^2)$$

or

$$a^4 - a^2(r^2 + c^2) + (\mathbf{r} \cdot \mathbf{c})^2 = 0.$$

13.7 If $\mathbf{a}$ is the position vector of a fixed point (x_1, y_1, z_1) and $\mathbf{r}$ the position vector of a variable point (x, y, z), describe the locus of $\mathbf{r}$ if
(a) $|(\mathbf{r}-\mathbf{a})| = 3$, (b) $(\mathbf{r}-\mathbf{a}) \cdot \mathbf{a} = 0$, (c) $(\mathbf{r}-\mathbf{a}) \cdot \mathbf{r} = 0$.

(a) The distance from $\mathbf{a}$ to $\mathbf{r}$ is a constant $(=3)$. Hence the locus of $\mathbf{r}$ is a spherical surface, centre $\mathbf{a}$, radius 3.
(b) $(\mathbf{r}-\mathbf{a})$ is perpendicular to $\mathbf{a}$. Hence the locus of $\mathbf{r}$ is a plane perpendicular to $\mathbf{a}$ and passing through (x_1, y_1, z_1).
(c) $(\mathbf{r}-\mathbf{a})$ is perpendicular to $\mathbf{r}$ (see Fig. 19). Hence the locus of $\mathbf{r}$ is a spherical surface with $\mathbf{a}$ as diameter (OQ), centre C $(\tfrac{1}{2}x_1, \tfrac{1}{2}y_1, \tfrac{1}{2}z_1)$, and radius $CQ = \tfrac{1}{2}\sqrt{x_1^2 + y_1^2 + z_1^2}$.

13.8 A plane containing the point $\mathbf{r}_0$ has its normal in the direction $\mathbf{n}$. A perpendicular is drawn to the plane from the point $\mathbf{r}_1$; find the point where the perpendicular meets the plane.

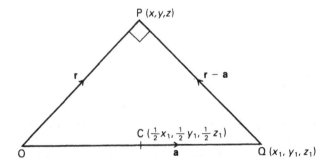

Fig. 19

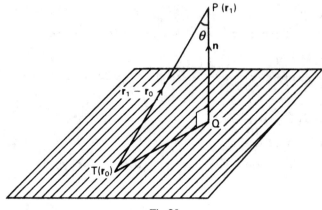

Fig. 20

If $\mathbf{r}$ is any point in the plane, then $\mathbf{r} - \mathbf{r}_0$ lies in the plane and therefore $(\mathbf{r} - \mathbf{r}_0) \cdot \mathbf{n} = 0$. Suppose P is the point $\mathbf{r}_1$ and Q is the projection of $\mathbf{r}_1$ in the plane (see Fig. 20). Then if T is the point $\mathbf{r}_0$

$$PQ = TP \cos \theta = (\overrightarrow{TP} \cdot n)/|\mathbf{n}|.$$

But $\overrightarrow{TP} = \mathbf{r}_1 - \mathbf{r}_0$. Hence

$$\mathbf{r}_Q = \mathbf{r}_1 - \frac{PQ\mathbf{n}}{|\mathbf{n}|} = \mathbf{r}_1 - \mathbf{n}\left[\frac{(\mathbf{r}_1 - \mathbf{r}_0) \cdot \mathbf{n}}{|\mathbf{n}|^2}\right].$$

13.9 Show that the two straight lines $\mathbf{r} = \mathbf{a} + h\mathbf{u}$ and $\mathbf{r} = \mathbf{b} + k\mathbf{v}$, where $\mathbf{a}$, $\mathbf{b}$ and $\mathbf{u}$, $\mathbf{v}$ are constant vectors, and h and k are scalar parameters, intersect if

$$\mathbf{v} \cdot (\mathbf{b} \wedge \mathbf{u}) = \mathbf{v} \cdot (\mathbf{a} \wedge \mathbf{u})$$

and that the point of intersection is

$$\mathbf{a} + \frac{(\mathbf{a} \cdot \mathbf{b} \wedge \mathbf{v})}{(\mathbf{v} \cdot \mathbf{a} \wedge \mathbf{u})} \mathbf{u}$$

or

$$\mathbf{b} + \frac{(\mathbf{a} \cdot \mathbf{b} \wedge \mathbf{u})}{(\mathbf{v} \cdot \mathbf{b} \wedge \mathbf{u})} \mathbf{v}.$$

The lines intersect if $\mathbf{a} + h\mathbf{u} = \mathbf{b} + k\mathbf{v}$. Taking the vector product with $\mathbf{u}$ gives $\mathbf{a} \wedge \mathbf{u} = \mathbf{b} \wedge \mathbf{u} + k\mathbf{v} \wedge \mathbf{u}$, since $\mathbf{u} \wedge \mathbf{u} = 0$. Now forming the scalar product with $\mathbf{v}$ gives the required result (using $\mathbf{v} \cdot (\mathbf{v} \wedge \mathbf{u}) = 0$)

$$\mathbf{v} \cdot (\mathbf{a} \wedge \mathbf{u}) = \mathbf{v} \cdot (\mathbf{b} \wedge \mathbf{u}).$$

Also

$$\mathbf{a} \cdot (\mathbf{a} \wedge \mathbf{u}) = \mathbf{a} \cdot (\mathbf{b} \wedge \mathbf{u}) + k\mathbf{a} \cdot (\mathbf{v} \wedge \mathbf{u}),$$

whence

$$k = -\frac{\mathbf{a} \cdot (\mathbf{b} \wedge \mathbf{u})}{\mathbf{a} \cdot (\mathbf{v} \wedge \mathbf{u})}, \quad \text{since} \quad \mathbf{a} \cdot (\mathbf{a} \wedge \mathbf{u}) = 0.$$

Hence the point of intersection is

$$\mathbf{r} = \mathbf{b} - \frac{\mathbf{a} \cdot (\mathbf{b} \wedge \mathbf{u})}{\mathbf{a} \cdot (\mathbf{v} \wedge \mathbf{u})} \mathbf{v} = \mathbf{b} + \frac{\mathbf{a} \cdot (\mathbf{b} \wedge \mathbf{u})}{\mathbf{v} \cdot (\mathbf{b} \wedge \mathbf{u})} \mathbf{v},$$

using the intersection condition and the cyclic properties of the scalar triple product $(\mathbf{a} \cdot (\mathbf{v} \wedge \mathbf{u}) = \mathbf{v} \cdot (\mathbf{u} \wedge \mathbf{a}) = \mathbf{u} \cdot (\mathbf{a} \wedge \mathbf{v})$. Likewise, solving for h, we find the other form for the point of intersection.

13.10 Find an expression for the length of the common perpendicular between the straight lines

$$\mathbf{r} = \mathbf{a} + \lambda \mathbf{u}, \qquad \mathbf{r} = \mathbf{b} + \mu \mathbf{v},$$

where $\mathbf{a}$, $\mathbf{b}$, $\mathbf{u}$, $\mathbf{v}$ are constant vectors, and λ, μ are scalar parameters. Hence, or otherwise, deduce that the two lines intersect if

$$(\mathbf{v} \wedge \mathbf{u}) \cdot (\mathbf{b} - \mathbf{a}) = 0.$$

Show that the length of the common perpendicular between the lines

$$\frac{x-1}{2} = \frac{y-3}{5} = \frac{z-1}{4}$$

$$\frac{x+1}{3} = \frac{y+2}{2} = \frac{z+1}{4}$$

is $\dfrac{22}{\sqrt{281}}$.

(L.U.)

Common perpendicular has direction $\mathbf{m} = \mathbf{u} \wedge \mathbf{v}$, since the line $\mathbf{v} = \mathbf{a} + \lambda \mathbf{u}$ has direction $\mathbf{u}$, and $\mathbf{u} \cdot (\mathbf{u} \wedge \mathbf{v}) = 0$ and likewise for the other line. The length p of the common perpendicular is therefore the projection of $(\mathbf{b} - \mathbf{a})$ on $\hat{\mathbf{m}}$:

$$p = |(\mathbf{b} - \mathbf{a}) \cdot \hat{\mathbf{m}}| = |(\mathbf{b} - \mathbf{a}) \cdot \mathbf{m}| / |\mathbf{m}|.$$

($\hat{\mathbf{m}}$ is the unit vector in the direction $\mathbf{m}$).

Hence the lines intersect if $p = 0$, which gives $(\mathbf{u} \wedge \mathbf{v}) \cdot (\mathbf{b} - \mathbf{a}) = 0$ as required. The first line given in Cartesian form has vector form $\mathbf{r} = \mathbf{a} + \lambda \mathbf{u}$, where $\mathbf{a} = (1, 3, 1)$ and $\mathbf{u} = (2, 5, 4)$. The second line has form $r = b + \mu v$, where $b = (-1, -2, -1)$ and $v = (3, 2, 4)$. Hence $m = 12i + 4j - 11k$, $|m| = \sqrt{281}$, and $b - a = -2i - 5j - 2k$. Hence $p = 22/\sqrt{281}$.

13.11 Which of the following sets of vectors is linearly dependent?
(a) $a = (0, 1, 1)$, $b = (1, 0, 1)$, $c = (1, 1, 0)$.
(b) $a = (1, 2, 3)$, $b = (4, 5, 6)$, $c = (7, 8, 9)$.

Linearly dependent if scalar triple product of vectors is zero (this corresponding geometrically to a zero volume of the parallelepiped whose sides are the given vectors, which implies that the three vectors lie in a plane).
(a) $a \wedge b = (1, 0, -1)$ and $(a \wedge b) \cdot c = 1$. Hence not linearly dependent.
(b) $a \wedge b = (-3, 6, -3)$ and $(a \wedge b) \cdot c = 0$. Hence linearly dependent. It is easily seen that

$$a = 2b - c.$$

13.12 Solve the vector equations for x:
(a) $x + (x \cdot b)a = c$,
   where a, b, c are given vectors, and $a \cdot b \neq -1$.
(b) $\alpha x + (a \wedge x) = b$,
   where a, b are given vectors and α is a non-zero constant.

(a) Consider the scalar product of each side of the equation with b.

$$b \cdot x + (x \cdot b)(b \cdot a) = b \cdot c.$$

Therefore $(x \cdot b)[1 + (a \cdot b)] = b \cdot c$ and hence $x \cdot b = \dfrac{b \cdot c}{1 + (a \cdot b)}$ provided $1 + a \cdot b \neq 0$ (which is given). Inserting this value of $x \cdot b$ into the equation gives the solution

$$x = c - a(x \cdot b) = c - a\frac{(b \cdot c)}{1 + (a \cdot b)}.$$

(b) Taking the vector product with a on both sides of the equation

$$\alpha(x \wedge a) + (a \wedge x) \wedge a = b \wedge a.$$

Then using the basic result $a \wedge (b \wedge c) = b(a \cdot c) - c(a \cdot b)$ to evaluate $(a \wedge x) \wedge a$, we find

$$\alpha(x \wedge a) + xa^2 - (a \cdot x)a = b \wedge a \qquad (a^2 = |a|^2).$$

Likewise taking the scalar product of each side of the equation with a gives

$$\alpha(x \cdot a) + (a \wedge x) \cdot a = b \cdot a$$

or, since $(\mathbf{a} \wedge \mathbf{x}) \cdot \mathbf{a} = 0$, $\mathbf{x} \cdot \mathbf{a} = \dfrac{\mathbf{b} \cdot \mathbf{a}}{\alpha}$. Inserting this form of $\mathbf{x} \cdot \mathbf{a}$ into $\alpha(\mathbf{x} \wedge \mathbf{a}) + \mathbf{x} a^2 - (\mathbf{a} \cdot \mathbf{x})\mathbf{a} = \mathbf{b} \wedge \mathbf{a}$, we have

$$\alpha(\mathbf{x} \wedge \mathbf{a}) = \mathbf{b} \wedge \mathbf{a} - \mathbf{x} a^2 + \mathbf{a}\left(\frac{\mathbf{b} \cdot \mathbf{a}}{\alpha}\right).$$

Inserting this form into the original equation leads to the solution

$$\mathbf{x} = \frac{\alpha^2 \mathbf{b} + \mathbf{a}(\mathbf{a} \cdot \mathbf{b}) + \alpha(\mathbf{b} \wedge \mathbf{a})}{\alpha(\alpha^2 + a^2)}.$$

13.13 Given that $x = A \cos(\omega t + \alpha)$, $y = B \cos(\omega t + \beta)$ and $z = Ct$, where A, B, C, ω, α, β are constants, show that the equation of motion of the point $\mathbf{r} = (x, y, z)$ is

$$\frac{d^2 \mathbf{r}}{dt^2} + \omega^2 \mathbf{r} = C\omega^2 t\mathbf{k}.$$

Now

$$\frac{d^2 x}{dt^2} = -\omega^2 x, \qquad \frac{d^2 y}{dt^2} = -\omega^2 y, \qquad \frac{d^2 z}{dt^2} = 0.$$

Hence

$$\frac{d^2 \mathbf{r}}{dt^2} = -\mathbf{i}\omega^2 x - \mathbf{j}\omega^2 y + 0$$

which is

$$\frac{d^2 \mathbf{r}}{dt^2} + \omega^2 \mathbf{r} = \omega^2 z\mathbf{k} = C\omega^2 t\mathbf{k}.$$

13.14 The position vector of a point at time t is given by

$$\mathbf{r} = (t + \sin t)\mathbf{i} + (t - \sin t)\mathbf{j} + (1 - \cos t)\sqrt{2}\mathbf{k}.$$

Show that $d\mathbf{r}/dt$ and $d^2\mathbf{r}/dt^2$ have constant magnitudes and are perpendicular vectors.

$$\dot{\mathbf{r}} = (1 + \cos t)\mathbf{i} + (1 - \cos t)\mathbf{j} + \sqrt{2}\sin t\,\mathbf{k},$$
$$\ddot{\mathbf{r}} = -\sin t\,\mathbf{i} + \sin t\,\mathbf{j} + \sqrt{2}\cos t\,\mathbf{k},$$
$$|\dot{\mathbf{r}}|^2 = (1 + \cos t)^2 + (1 - \cos t)^2 + 2\sin^2 t = 4,$$
$$|\ddot{\mathbf{r}}|^2 = \sin^2 t + \sin^2 t + 2\cos^2 t = 2.$$

Hence $\dot{\mathbf{r}}$ and $\ddot{\mathbf{r}}$ are constant-magnitude vectors. Also it is easily seen that $\dot{\mathbf{r}} \cdot \ddot{\mathbf{r}} = 0$. Hence $\dot{\mathbf{r}}$ and $\ddot{\mathbf{r}}$ are perpendicular.

14. MATRIX ALGEBRA

[Transformations, mappings, groups, inverse matrices, solution of equations, Gaussian elimination, LU-decomposition, symmetric and skew-symmetric matrices, Hermitian and skew-Hermitian matrices, orthogonal and unitary matrices, eigenvalues and eigenvectors, diagonalisation, quadratic forms]

14.1   In Fig. 21 consider the two transformations R and S, where R is a rotation anti-clockwise about O through $2\pi/3$, and S is a reflection in the x-axis. Show that if P is some point in the plane

$$RS(P) \neq SR(P).$$

$RS(P)$ and $SR(P)$ are shown in Fig. 21. The result of reflecting first and then rotating is not the same as first rotating and then reflecting. The two transformations do not commute.

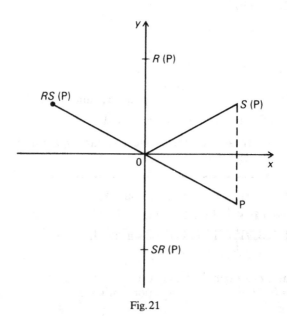

Fig. 21

14.2 If f is the mapping which sends x into y, and g the mapping which sends y into z, show that

$$(gf)^{-1} \equiv f^{-1}g^{-1},$$

where f^{-1}, g^{-1} are the inverse mappings of f and g respectively.

$y = f(x)$, $z = g(y)$. Therefore $x = f^{-1}(y)$, $y = g^{-1}(z)$. Similarly $z = gf(x)$, and hence $x = (gf)^{-1}(z)$. Therefore $(gf)^{-1}(z) = f^{-1}g^{-1}(z)$, or $(gf)^{-1} \equiv f^{-1}g^{-1}$.

14.3 Show that the set of four matrices

$$\begin{pmatrix} 0 & 1 \\ -1 & 0 \end{pmatrix}, \begin{pmatrix} -1 & 0 \\ 0 & -1 \end{pmatrix}, \begin{pmatrix} 0 & -1 \\ 1 & 0 \end{pmatrix}, \begin{pmatrix} 1 & 0 \\ 0 & 1 \end{pmatrix}$$

forms an Abelian group of order four.

Product of any two matrices in the set is another matrix in the set; for example,

$$\begin{pmatrix} 0 & 1 \\ -1 & 0 \end{pmatrix}\begin{pmatrix} 0 & -1 \\ 1 & 0 \end{pmatrix} = \begin{pmatrix} 1 & 0 \\ 0 & 1 \end{pmatrix}.$$

Inverse of any matrix in the set is another matrix in the set; for example,

$$\begin{pmatrix} 0 & 1 \\ -1 & 0 \end{pmatrix}^{-1} = \begin{pmatrix} 0 & -1 \\ 1 & 0 \end{pmatrix}.$$

The matrix $\begin{pmatrix} 1 & 0 \\ 0 & 1 \end{pmatrix}$ corresponds to the unit element of the group since its product with every other matrix in the set leaves each matrix unchanged.

The associative law for group elements is true for matrix multiplication (e.g. if A, B, C are matrices then $A(BC) = (AB)C = ABC$). Hence the set of four matrices forms a group with four elements (i.e. a group of order four). This group is Abelian since all elements commute.

14.4 Given $A = \begin{pmatrix} 1 & 1 \\ 0 & 1 \end{pmatrix}$ and $B = \begin{pmatrix} 1 & 1 \\ 1 & 0 \end{pmatrix}$, show that

$$(A+B)(A-B) = \begin{pmatrix} -2 & 2 \\ -1 & 1 \end{pmatrix}.$$

and verify that $(A+B)(A-B) \neq A^2 - B^2$.

$$A + B = \begin{pmatrix} 2 & 2 \\ 1 & 1 \end{pmatrix}, \quad A - B = \begin{pmatrix} 0 & 0 \\ -1 & 1 \end{pmatrix},$$

$$(A + B)(A - B) = \begin{pmatrix} 2 & 2 \\ 1 & 1 \end{pmatrix}\begin{pmatrix} 0 & 0 \\ -1 & 1 \end{pmatrix} = \begin{pmatrix} -2 & 2 \\ -1 & 1 \end{pmatrix}.$$

$$A^2 = \begin{pmatrix} 1 & 2 \\ 0 & 1 \end{pmatrix}, \quad B^2 = \begin{pmatrix} 2 & 1 \\ 1 & 1 \end{pmatrix},$$

$$A^2 - B^2 = \begin{pmatrix} -1 & 1 \\ -1 & 0 \end{pmatrix} \neq (A + B)(A - B).$$

$(A + B)(A - B) = A^2 + BA - AB - B^2 \neq A^2 - B^2$ since $BA \neq AB$.

14.5 Show that if A and B are square matrices satisfying the equation $A^2 = B$, then A and B commute. Hence find matrices A such that

$$A^2 = \begin{pmatrix} 1 & 1 & 0 \\ 0 & 1 & 0 \\ 2 & 0 & 1 \end{pmatrix}.$$

If $A^2 = B$ then $A^3 = AB$, and $A^3 = BA$. therefore $AB = BA$. Since

$$A^2 = B = \begin{pmatrix} 1 & 1 & 0 \\ 0 & 1 & 0 \\ 2 & 0 & 1 \end{pmatrix}$$

consider

$$A = \begin{pmatrix} a_{11} & a_{12} & a_{13} \\ a_{21} & a_{22} & a_{23} \\ a_{31} & a_{32} & a_{33} \end{pmatrix}.$$

Then

$$\begin{pmatrix} 1 & 1 & 0 \\ 0 & 1 & 0 \\ 2 & 0 & 1 \end{pmatrix}\begin{pmatrix} a_{11} & a_{12} & a_{13} \\ a_{21} & a_{22} & a_{23} \\ a_{31} & a_{32} & a_{33} \end{pmatrix} = \begin{pmatrix} a_{11} & a_{12} & a_{13} \\ a_{21} & a_{22} & a_{23} \\ a_{31} & a_{32} & a_{33} \end{pmatrix}\begin{pmatrix} 1 & 1 & 0 \\ 0 & 1 & 0 \\ 2 & 0 & 1 \end{pmatrix},$$

whence comparing elements on each side (after some algebraic manipulation) we have

$$A = \pm\begin{pmatrix} 1 & \frac{1}{2} & 0 \\ 0 & 1 & 0 \\ 1 & -\frac{1}{4} & 1 \end{pmatrix}.$$

14.6 Given that A is an $(m \times n)$ matrix with $A^{T}A$ non-singular

(A^T being the transpose of A) and

$$B = I - A(A^T A)^{-1} A^T,$$

show that $B^2 = B$, where I is the unit matrix of order m.

$$B = [I - A(A^T A)^{-1} A^T][I - A(A^T A)^{-1} A^T]$$
$$= I^2 - 2A(A^T A)^{-1} A^T + A(A^T A)^{-1} A^T A(A^T A)^{-1} A^T$$
$$= I - 2A(A^T A)^{-1} A^T + A(A^T A)^{-1} A^T \text{ since } (A^T A)^{-1} A^T A = I.$$

Therefore $B^2 = B$.

14.7 Solve the equations

$$5x + 3y + z = 31,$$
$$2x + 6y + 3z = 32,$$
$$x + 3y + 7z = 27,$$

(a) by matrix inversion, (b) by Gaussian elimination, (c) by LU-decomposition.

(a) We write the equations as $AX = H$, where

$$A = \begin{pmatrix} 5 & 3 & 1 \\ 2 & 6 & 3 \\ 1 & 3 & 7 \end{pmatrix}, \quad X = \begin{pmatrix} x \\ y \\ z \end{pmatrix} \text{ and } H = \begin{pmatrix} 31 \\ 32 \\ 27 \end{pmatrix}.$$

Then, provided $|A| \neq 0$, A^{-1} exists and the solution may be written as $X = A^{-1} H$. The inverse of A is (generally) $A^{-1} = \text{adj } A/|A|$, where adj A is the adjoint matrix of A, defined as the transposed matrix of the cofactors of A. For any square matrix A with elements a_{ik} the cofactor A_{ik} of the element a_{ik} is defined as $(-1)^{i+k} \times$ the determinant remaining when the row and column containing the element a_{ik} are deleted. Hence taking the specific form of A in this problem we see that the cofactor of the element in the first row and first column (a_{11}) is $(-1)^{1+1}(6.7 - 3.3) = 33$, and that the cofactor of the element in the first row and second column (a_{12}) is $(-1)^{1+2}(2.7 - 3.1) = -11$. Similarly for the other elements. The matrix of cofactors of A is found to be

$$\begin{pmatrix} 33 & -11 & 0 \\ -18 & 34 & -12 \\ 3 & -13 & 24 \end{pmatrix}$$

and accordingly

$$\text{adj } A = \begin{pmatrix} 33 & -18 & 3 \\ -11 & 34 & -13 \\ 0 & -12 & 24 \end{pmatrix}.$$

Direct evaluation of the determinant of A $(|A|)$ gives 132. Hence

$$A^{-1} = \tfrac{1}{132} \begin{pmatrix} 33 & -18 & 3 \\ -11 & 34 & -13 \\ 0 & -12 & 24 \end{pmatrix}.$$

Accordingly

$$X = \tfrac{1}{132} \begin{pmatrix} 33 & -18 & 3 \\ -11 & 34 & -13 \\ 0 & -12 & 24 \end{pmatrix} \begin{pmatrix} 31 \\ 32 \\ 27 \end{pmatrix} = \begin{pmatrix} 4 \\ 3 \\ 2 \end{pmatrix},$$

and the solutions are therefore $x = 4$, $y = 3$, $z = 2$.

(b) Subtracting twice the third equation from the second equation we get

$$5x + 3y + z = 31,$$
$$0x + 0y - 11z = -22,$$
$$x + 3y + 7z = 27,$$

whence directly $z = 2$. Hence the remaining equations are

$$5x + 3y = 29,$$
$$x + 3y = 13,$$

and therefore $x = 4$ and $y = 3$.

(c) Let L be a lower triangular matrix with the form

$$L = \begin{pmatrix} 1 & 0 & 0 \\ l_{21} & 1 & 0 \\ l_{31} & l_{32} & 1 \end{pmatrix}$$

and U be an upper triangular matrix with the form

$$U = \begin{pmatrix} u_{11} & u_{12} & u_{13} \\ 0 & u_{22} & u_{23} \\ 0 & 0 & u_{33} \end{pmatrix}.$$

Then the L and the U may be determined so that $A = LU$. The equation $AX = H$ may now be written as $LUX = H$. Writing $UX = Y$ we first solve $LY = H$ for Y, and then inserting the values of Y, solve $UX = Y$ for X.

A can be written in the appropriate form as

$$A = LU = \begin{pmatrix} 1 & 0 & 0 \\ 0.4 & 1 & 0 \\ 0.2 & 0.5 & 1 \end{pmatrix} \begin{pmatrix} 5 & 3 & 1 \\ 0 & 4.8 & 2.6 \\ 0 & 0 & 5.5 \end{pmatrix},$$

and hence $LY = H$ is

$$\begin{pmatrix} 1 & 0 & 0 \\ 0.4 & 1 & 0 \\ 0.2 & 0.5 & 1 \end{pmatrix} \begin{pmatrix} y_1 \\ y_2 \\ y_3 \end{pmatrix} = \begin{pmatrix} 31 \\ 32 \\ 27 \end{pmatrix},$$

(where y_1, y_2, y_3 are the elements of Y). Reading from the first row, $y_1 = 31$. Hence, from the second row, $0.4y_1 + y_2 = 32$ or $y_2 = 19.6$, and finally, from the third row, $y_3 = 11$. $UX = Y$ is therefore

$$\begin{pmatrix} 5 & 3 & 1 \\ 0 & 4.8 & 2.6 \\ 0 & 0 & 5.5 \end{pmatrix} \begin{pmatrix} x \\ y \\ z \end{pmatrix} = \begin{pmatrix} 31 \\ 19.6 \\ 11 \end{pmatrix}$$

and reading from the third row gives $z = 2$. Hence from the second row $4.8y + 2.6z = 19.6$ or $y = 3$, and finally, from the first row, $x = 4$.

14.8 Given a symmetric matrix

$$A = \begin{pmatrix} 1 & 2 & 6 \\ 2 & 5 & 15 \\ 6 & 15 & 46 \end{pmatrix},$$

determine a lower triangular matrix L such that $LL^T = A$. Hence obtain the solution x of the equation

$$Ax = b = \begin{pmatrix} 1 \\ 2 \\ 3 \end{pmatrix}$$

by solving $Ly = b$ for y, and then $L^T x = y$ for x.   (L.U.)

$$L = \begin{pmatrix} a_{11} & 0 & 0 \\ a_{21} & a_{22} & 0 \\ a_{31} & a_{32} & a_{33} \end{pmatrix}, \qquad L^T = \begin{pmatrix} a_{11} & a_{21} & a_{31} \\ 0 & a_{22} & a_{32} \\ 0 & 0 & a_{33} \end{pmatrix}.$$

$LL^T = A$ gives $a_{11}^2 = 1$, therefore $a_{11} = 1$; $a_{11}a_{21} = 2$, therefore $a_{21} = 2$; $a_{11}a_{31} = 6$, therefore $a_{31} = 6$; $a_{21}^2 + a_{22}^2 = 5$, therefore $a_{22} = 1$; $a_{21}a_{31} + a_{22}a_{32} = 15$, therefore $a_{32} = 3$; $a_{31}^2 + a_{32}^2 + a_{33}^2 = 46$, therefore $a_{33} = 1$. Hence

$$\begin{pmatrix} 1 & 0 & 0 \\ 2 & 1 & 0 \\ 6 & 3 & 1 \end{pmatrix} \begin{pmatrix} y_1 \\ y_2 \\ y_3 \end{pmatrix} = \begin{pmatrix} 1 \\ 2 \\ 3 \end{pmatrix}$$

(where y_1, y_2, y_3 are the elements of y). Reading downwards from the first row $y_1 = 1$; $2y_1 + y_2 = 2$, therefore $y_2 = 0$; $6y_1 + 3y_2 + y_3 = 3$, therefore $y_3 = -3$. Finally

$$\begin{pmatrix} 1 & 2 & 6 \\ 0 & 1 & 3 \\ 0 & 0 & 1 \end{pmatrix}\begin{pmatrix} x_1 \\ x_2 \\ x_3 \end{pmatrix} = \begin{pmatrix} 1 \\ 0 \\ -3 \end{pmatrix}$$

(where x_1, x_2, x_3 are the elements of x). Reading upwards from the third row in the same way we find

$$x = \begin{pmatrix} 1 \\ 9 \\ -3 \end{pmatrix}.$$

14.9 Show, by the method of induction (see 2.5), that

$$S^n = \begin{pmatrix} t^n & nt^{n-1} & \tfrac{1}{2}n(n-1)t^{n-2} \\ 0 & t^n & nt^{n-1} \\ 0 & 0 & t^n \end{pmatrix},$$

where

$$S = \begin{pmatrix} t & 1 & 0 \\ 0 & t & 1 \\ 0 & 0 & t \end{pmatrix}.$$

Consider

$$S^2 = \begin{pmatrix} t & 1 & 0 \\ 0 & t & 1 \\ 0 & 0 & t \end{pmatrix}\begin{pmatrix} t & 1 & 0 \\ 0 & t & 1 \\ 0 & 0 & t \end{pmatrix} = \begin{pmatrix} t^2 & 2t & 1 \\ 0 & t^2 & 2t \\ 0 & 0 & t^2 \end{pmatrix}.$$

Hence the formula is true for $n = 2$.
Now assume the result is true for some $n = k$ (say). Then

$$S^{k+1} = S \cdot S^k = \begin{pmatrix} t & 1 & 0 \\ 0 & t & 1 \\ 0 & 0 & t \end{pmatrix}\begin{pmatrix} t^k & kt^{k-1} & \tfrac{1}{2}k(k-1)t^{k-2} \\ 0 & t^k & kt^{k-1} \\ 0 & 0 & t^k \end{pmatrix}$$

$$= \begin{pmatrix} t^{k+1} & (k+1)t^k & \tfrac{1}{2}k(k+1)t^{k-1} \\ 0 & t^{k+1} & (k+1)t^k \\ 0 & 0 & t^{k+1} \end{pmatrix},$$

which is the formula for $n = k + 1$. Hence if true for $n = k$, then true also for $n = k + 1$. But, since the result is true for $n = 2$, it is true therefore for $n = 3, 4, \ldots$.

14.10 (a) Show that the general matrix

$$A = \begin{pmatrix} a_{11} & a_{12} \\ a_{21} & a_{22} \end{pmatrix}$$

may be written in the form $A = B + C$, where B is a symmetric matrix and C is a skew-symmetric matrix. Show that B^2 is a symmetric matrix and that C^2 is a diagonal matrix.

(b) Given $E = \begin{pmatrix} a & b \\ c & d \end{pmatrix}$, where $b \neq 0$, and if $E^2 = I$, evaluate c and d in terms of a and b. State briefly why this matrix E is of interest.

(L.U.)

(a) Let

$$B = \begin{pmatrix} a_{11} & x \\ x & a_{22} \end{pmatrix}, \qquad C = \begin{pmatrix} 0 & y \\ -y & 0 \end{pmatrix}.$$

Then $x + y = a_{12}$, $x - y = a_{21}$ give $x = \frac{1}{2}(a_{12} + a_{21})$, $y = \frac{1}{2}(a_{12} - a_{21})$. Hence

$$B^2 = \begin{pmatrix} a_{11} & x \\ x & a_{22} \end{pmatrix}\begin{pmatrix} a_{11} & x \\ x & a_{22} \end{pmatrix} = \begin{pmatrix} a_{11}^2 + x^2 & a_{11}x + a_{22}x \\ a_{11}x + a_{22}x & x^2 + a_{22}^2 \end{pmatrix},$$

which is symmetric. Also

$$C^2 = \begin{pmatrix} -y^2 & 0 \\ 0 & y^2 \end{pmatrix},$$

which is diagonal.
(b) $E^2 = I$ gives

$$\begin{pmatrix} a^2 + bc & ab + db \\ ac + dc & d^2 + bc \end{pmatrix} = \begin{pmatrix} 1 & 0 \\ 0 & 1 \end{pmatrix}.$$

Hence if $b \neq 0$, $d = -a$, $c = (1 - a^2)/b$. Hence

$$E = \begin{pmatrix} a & b \\ (1-a^2)/b & -a \end{pmatrix}$$

is a square root of I. Clearly there are *infinitely* many matrices which are square roots of I.

14.11 Verify that if

$$A = \begin{pmatrix} 1 & 2 \\ 4 & 3 \end{pmatrix},$$

then $A^2 - 4A - 5I = 0$, and hence evaluate A^{-1}.

$$A^2 = \begin{pmatrix} 1 & 2 \\ 4 & 3 \end{pmatrix}\begin{pmatrix} 1 & 2 \\ 4 & 3 \end{pmatrix} = \begin{pmatrix} 9 & 8 \\ 16 & 17 \end{pmatrix}.$$

$$A^2 - 4A - 5I = \begin{pmatrix} 9 & 8 \\ 16 & 17 \end{pmatrix} - 4\begin{pmatrix} 1 & 2 \\ 4 & 3 \end{pmatrix} - 5\begin{pmatrix} 1 & 0 \\ 0 & 1 \end{pmatrix} = \begin{pmatrix} 0 & 0 \\ 0 & 0 \end{pmatrix}$$

$A^{-1}(A^2 - 4A - 5I) = A^{-1}0 = 0$. Therefore $A - 4I - 5A^{-1} = 0$. Hence

$$A^{-1} = \tfrac{1}{5}(A - 4I) = \frac{1}{5}\left[\begin{pmatrix} 1 & 2 \\ 4 & 3 \end{pmatrix} - 4\begin{pmatrix} 1 & 0 \\ 0 & 1 \end{pmatrix}\right] = \frac{1}{5}\begin{pmatrix} -3 & 2 \\ 4 & -1 \end{pmatrix}.$$

14.12 If $A = \begin{pmatrix} 3 & 2 \\ 2 & 3 \end{pmatrix}$, $S = \dfrac{1}{\sqrt{2}}\begin{pmatrix} 1 & 1 \\ 1 & -1 \end{pmatrix}$, prove that $D = S^{-1}AS$ is a diagonal matrix. Verify that $\mathrm{Tr}(S^{-1}AS) = \mathrm{Tr}\,A$, and that $\det A = \det(S^{-1}AS)$, (Tr denotes trace, the sum of the diagonal elements). Show also that $D^n = S^{-1}A^nS$, where n is a positive integer.

$$S^{-1} = -\frac{1}{\sqrt{2}}\begin{pmatrix} -1 & -1 \\ -1 & 1 \end{pmatrix}.$$

$$S^{-1}AS = -\frac{1}{\sqrt{2}}\begin{pmatrix} -1 & -1 \\ -1 & 1 \end{pmatrix}\begin{pmatrix} 3 & 2 \\ 2 & 3 \end{pmatrix}\frac{1}{\sqrt{2}}\begin{pmatrix} 1 & 1 \\ 1 & -1 \end{pmatrix} = \begin{pmatrix} 5 & 0 \\ 0 & 1 \end{pmatrix} = D.$$

$\mathrm{Tr}\,D = 5 + 1 = 6$, $\mathrm{Tr}\,A = 3 + 3 = 6$.

$\det D = 5$, $\det A = 3^2 - 2^2 = 5$.

$D^2 = S^{-1}AS \cdot S^{-1}AS = S^{-1}A^2S$, since $SS^{-1} = I$. Similarly $D^n = S^{-1}A^nS$, where n is a positive integer.

14.13 Express the quadratic forms (a) $2x^2 - 5xy + 5y^2$, (b) $x^2 - 2xy + 2y^2 - 2yz + 2z^2$ in the form $U^T A U$, where U is a suitable column vector and A is a symmetric matrix.

(a) $U = \begin{pmatrix} x \\ y \end{pmatrix}$, $U^T = (x \quad y)$, $A = \begin{pmatrix} 2 & -\frac{5}{2} \\ -\frac{5}{2} & 5 \end{pmatrix}$.

(b) $U = \begin{pmatrix} x \\ y \\ z \end{pmatrix}$, $U^T = (x \quad y \quad z)$, $A = \begin{pmatrix} 1 & -1 & 0 \\ -1 & 2 & -1 \\ 0 & -1 & 2 \end{pmatrix}$.

14.14 Show that the matrix

$$\frac{1}{15}\begin{pmatrix} 5 & -14 & 2 \\ -10 & -5 & -10 \\ 10 & 2 & -11 \end{pmatrix}$$

is orthogonal and verify that its rows and columns form orthonormal sets of vectors.

Orthogonal if $A^T = A^{-1}$, or $AA^T = A^TA = I$.

$$A^T = \frac{1}{15}\begin{pmatrix} 5 & -10 & 10 \\ -14 & -5 & 2 \\ 2 & -10 & -11 \end{pmatrix}.$$

$$AA^T = \frac{1}{15}\begin{pmatrix} 5 & -14 & 2 \\ -10 & -5 & -10 \\ 10 & 2 & -11 \end{pmatrix} \times \frac{1}{15}\begin{pmatrix} 5 & -10 & 10 \\ -14 & -5 & 2 \\ 2 & -10 & -11 \end{pmatrix}$$

$$= \begin{pmatrix} 1 & 0 & 0 \\ 0 & 1 & 0 \\ 0 & 0 & 1 \end{pmatrix} = I.$$

Likewise $A^TA = I$. Hence A is orthogonal. Let

$$X_1 = \frac{1}{15}\begin{pmatrix} 5 \\ -10 \\ 10 \end{pmatrix}, \quad X_2 = \frac{1}{15}\begin{pmatrix} -14 \\ -5 \\ 2 \end{pmatrix}, \quad X_3 = \frac{1}{15}\begin{pmatrix} 2 \\ -10 \\ -11 \end{pmatrix}.$$

Then

$$X_1^T X_1 = \frac{1}{225}(5 \quad -10 \quad 10)\begin{pmatrix} 5 \\ -10 \\ 10 \end{pmatrix} = 1,$$

$$X_1^T X_2 = \frac{1}{225}(5 \quad -10 \quad 10)\begin{pmatrix} -14 \\ -5 \\ 2 \end{pmatrix} = 0, \quad X_1^T X_3 = 0,$$

and so on. Hence finally

$$X_i^T X_j = \delta_{ij} = \begin{cases} 0, & i \neq j, \\ 1, & i = j, \end{cases} \quad (i, j = 1, 2, 3)$$

where δ_{ij} is the Kronecker delta symbol. Similarly for the rows. Hence the rows and columns form sets of orthogonal unit vectors (or *orthonormal* vectors). (N.B. This is a property of all orthogonal matrices.)

14.15 (a) Prove that if A is skew-symmetric, then

$$(I - A)(I + A)^{-1}$$

is orthogonal (assuming that $I + A$ is non-singular).
(b) Prove that if A is skew-Hermitian, then

$$(I - A)(I + A)^{-1}$$

is unitary (assuming that $I + A$ is non-singular).

(a) Let $(I-A)(I+A)^{-1} = U$. Then U is orthogonal if $UU^T = U^TU = I$. Now using the rule for the transpose of a matrix product (i.e. $(AB)^T = B^T A^T$) we find

$$U^T = (I + A^T)^{-1}(I - A^T).$$

Since A is skew-symmetric, $A^T = -A$, and hence

$$U^T = (I - A)^{-1}(I + A).$$

Then

$$U^TU = (I-A)^{-1}(I+A)(I-A)(I+A)^{-1}$$
$$= (I-A)^{-1}(I-A)(I+A)(I+A)^{-1} = I.$$

Likewise $UU^T = I$. Hence U is orthogonal.

(b) Let $(I-A)(I+A)^{-1} = U$. Then U is unitary if $U^\dagger U = UU^\dagger = I$, where $U^\dagger$ $(=U^{*T})$ is the Hermitian conjugate of U. Now $U^\dagger = (I+A^\dagger)^{-1}(I-A^\dagger)$. Since A is skew-Hermitian, $A^\dagger = -A$, and hence $U^\dagger = (I-A)^{-1}(I+A)$. Then

$$U^\dagger U = (I-A)^{-1}(I+A)(I-A)(I+A)^{-1}$$
$$= (I-A)^{-1}(I-A)(I+A)(I+A)^{-1} = I.$$

Similarly $UU^\dagger = I$. Hence U is unitary.

14.16 Find eigenvalues and eigenvectors (normalized to unit length) of the matrix

$$A = \begin{pmatrix} 2 & 3 & -1 \\ 0 & -4 & 2 \\ 0 & -5 & 3 \end{pmatrix}.$$

$|A - \lambda I| = 0$ is the characteristic equation for the eigenvalues λ. Hence

$$\begin{vmatrix} 2-\lambda & 3 & -1 \\ 0 & -4-\lambda & 2 \\ 0 & -5 & 3-\lambda \end{vmatrix} = 0$$

or $(2-\lambda)[-(4+\lambda)(3-\lambda)+10] = 0$. Therefore $(2-\lambda)(\lambda-1)(\lambda+2) = 0$ or $\lambda = 2, -2, 1$.

The eigenvector equation is $AX = \lambda X$, where X is an eigenvector of A. When $\lambda = 2$, $AX = 2X$ or $(A-2I)X = 0$ (where 0 is the zero matrix). Hence

$$\begin{pmatrix} 0 & 3 & -1 \\ 0 & -6 & 2 \\ 0 & -5 & 1 \end{pmatrix} \begin{pmatrix} x_1 \\ x_2 \\ x_3 \end{pmatrix} = \begin{pmatrix} 0 \\ 0 \\ 0 \end{pmatrix}$$

(where x_1, x_2, x_3 are the elements of X). It follows that $3x_2 - x_3 = 0$, $-6x_2 + 2x_3 = 0$, $-5x_2 + x_3 = 0$, giving $x_2 = x_3 = 0$, x_1 arbitrary. Therefore

$$X^{(1)} = \begin{pmatrix} \alpha \\ 0 \\ 0 \end{pmatrix} \quad (\alpha \text{ arbitrary}).$$

Normalizing to unit length by requiring $X^{(1)\mathrm{T}}X^{(1)} = 1$ gives finally

$$X^{(1)} = \begin{pmatrix} 1 \\ 0 \\ 0 \end{pmatrix}.$$

Similarly for $\lambda = -2$, 1, respectively

$$X^{(2)} = \frac{1}{3}\begin{pmatrix} 1 \\ -2 \\ -2 \end{pmatrix}, \quad X^{(3)} = \frac{1}{\sqrt{30}}\begin{pmatrix} 1 \\ -2 \\ -5 \end{pmatrix}.$$

14.17 Show that if $(A - \lambda I)X = 0$, then $(T^{-1}AT - \lambda I)T^{-1}X = 0$.

Now $(T^{-1}AT - \lambda I)T^{-1}X = T^{-1}ATT^{-1}X - \lambda IT^{-1}X$
$$= T^{-1}(A - \lambda I)X,$$

since $TT^{-1} = I$. Therefore since $(A - \lambda I)X = 0$, the result follows.

14.18 Find eigenvectors of the matrix

$$\begin{pmatrix} 2 & 1 & 2 \\ 0 & 2 & 3 \\ 0 & 0 & 5 \end{pmatrix}$$

and show that only two of them are linearly independent.

Characteristic equation is

$$|A - \lambda I| = \begin{vmatrix} 2-\lambda & 1 & 2 \\ 0 & 2-\lambda & 3 \\ 0 & 0 & 5-\lambda \end{vmatrix} = 0,$$

whence the eigenvalues are found to be $\lambda = 2, 2, 5$, two of which are identical. Using the eigenvector equation $AX = \lambda X$, we have for $\lambda = 2$

$$AX^{(1)} = 2X^{(1)} \quad \text{or} \quad \begin{pmatrix} 0 & 1 & 2 \\ 0 & 0 & 3 \\ 0 & 0 & 3 \end{pmatrix}\begin{pmatrix} x_1 \\ x_2 \\ x_3 \end{pmatrix} = \begin{pmatrix} 0 \\ 0 \\ 0 \end{pmatrix}.$$

Therefore $x_2 + 2x_3 = 0$, $x_3 = 0$ or $x_2 = x_3 = 0$, x_1 arbitrary. Hence the eigenvectors for the two repeated eigenvalues are (in normalized form)

$$X^{(1)} = X^{(2)} = \begin{pmatrix} 1 \\ 0 \\ 0 \end{pmatrix}.$$

When $\lambda = 5$, $AX^{(3)} = 5X^{(3)}$, and

$$X^{(3)} = \frac{1}{\sqrt{3}} \begin{pmatrix} 1 \\ 1 \\ 1 \end{pmatrix}.$$

Hence $X^{(3)}$ and $X^{(1)}$ (or $X^{(2)}$) are linearly independent.

14.19 Diagonalize the matrix

$$A = \begin{pmatrix} 2 & 3 & -1 \\ 0 & -4 & 2 \\ 0 & -5 & 3 \end{pmatrix}$$

and hence derive A^5.

Using the results of 14.16, the matrix of eigenvectors may be written as

$$S = \begin{pmatrix} 1 & \dfrac{1}{3} & \dfrac{1}{\sqrt{30}} \\ 0 & -\dfrac{2}{3} & -\dfrac{2}{\sqrt{30}} \\ 0 & -\dfrac{2}{3} & -\dfrac{5}{\sqrt{30}} \end{pmatrix}.$$

Evaluating S^{-1} gives

$$\begin{pmatrix} 1 & \dfrac{1}{2} & 0 \\ 0 & -\dfrac{5}{2} & 1 \\ 0 & \dfrac{\sqrt{30}}{3} & -\dfrac{\sqrt{30}}{3} \end{pmatrix}.$$

Hence

$$D = S^{-1}AS = \begin{pmatrix} 2 & 0 & 0 \\ 0 & -2 & 0 \\ 0 & 0 & 1 \end{pmatrix} \quad \text{and} \quad D^5 = \begin{pmatrix} 2^5 & 0 & 0 \\ 0 & (-2)^5 & 0 \\ 0 & 0 & 1 \end{pmatrix}.$$

Now $D^5 = S^{-1}A^5S$ and hence $A^5 = SD^5S^{-1}$. Finally

$$A^5 = \begin{pmatrix} 32 & 43 & -11 \\ 0 & -54 & 22 \\ 0 & -55 & 23 \end{pmatrix}.$$

14.20 Prove that a (2×2) Hermitian matrix must have two different eigenvalues unless it is a multiple of the unit matrix.

Let

$$H = \begin{pmatrix} a & c+id \\ c-id & b \end{pmatrix},$$

where a, b, c, d are real constants. Then H is Hermitian since $H = H^\dagger$ (N.B. $H^\dagger = (H^*)^T$). The characteristic equation for the eigenvalues of H is $|H - \lambda I| = 0$, which is

$$\begin{vmatrix} a-\lambda & c+id \\ c-id & b-\lambda \end{vmatrix} = 0,$$

which gives the quadratic $\lambda^2 - (a+b)\lambda + ab - c^2 - d^2$ for the two values of λ. Solving we find

$$\lambda = \tfrac{1}{2}(a+b) \pm \tfrac{1}{2}\sqrt{(a-b)^2 + 4(c^2 + d^2)}.$$

The two roots are necessarily different unless $(a-b)^2 + 4(c^2 + d^2) = 0$. This expression is zero only when $a = b$, $c = d = 0$, since it is the sum of squared terms. In this case

$$H = \begin{pmatrix} a & 0 \\ 0 & a \end{pmatrix} = a \begin{pmatrix} 1 & 0 \\ 0 & 1 \end{pmatrix},$$

which is a multiple of the unit matrix.

15. LINE AND DOUBLE INTEGRALS

[Line integrals: parametric form, integrals around closed paths, path independence, vector fields. Double integrals: reversal of order of integration, transformation of variables, Green's theorem in the plane, convolution integral]

15.1 Evaluate $\int_C [(x^2+y^2)\,dx - 2xy\,dy]$ from the origin $(0,0)$ to $(1,1)$ along the paths (a) $y=x$, (b) $y=\sqrt{x}$.

(a) In Fig. 22, C_1 is the line $y=x$. Hence on C_1 $dy=dx$.

$$\int_{C_1} = \int_{x=0}^{x=1} (x^2+x^2)\,dx - 2x^2\,dx = \left(\frac{2x^3}{3} - \frac{2x^3}{3}\right)_0^1 = 0.$$

(b) In Fig. 22, C_2 is the curve $y=\sqrt{x}$. Hence on C_2 $dy = \frac{1}{2\sqrt{x}}dx$.

$$\int_{C_2} = \int_{x=0}^{x=1} (x^2+x)\,dx - 2x^{3/2}\frac{1}{2\sqrt{x}}dx$$
$$= \int_0^1 (x^2+x-x)\,dx = \left(\frac{x^3}{3}\right)_0^1 = \frac{1}{3}.$$

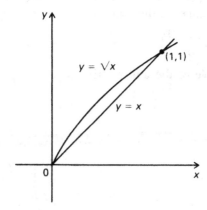

Fig. 22

15.2 Evaluate

$$\int_{(0,0)}^{(1,1)} (y^2 + kxy)\,dx + (x^2 + kxy)\,dy,$$

where k is an arbitrary constant, along (a) $y = x$, (b) $y = x^2$, and find that value of k which makes the integral independent of path.

(a) $\displaystyle\int_{x=0}^{x=1} (x^2 + kx^2)\,dx + (x^2 + kx^2)\,dx = 2\left(\frac{x^3}{3} + \frac{kx^3}{3}\right)_0^1 = \frac{2}{3}(1+k).$

(b) $\displaystyle\int_{x=0}^{x=1} (x^4 + kx^3)\,dx + (x^2 + kx^3)2x\,dx = \frac{1}{5} + \frac{k}{4} + 2\left(\frac{1}{4} + \frac{k}{5}\right)$

$$= \frac{13k}{20} + \frac{7}{10}.$$

A line integral of the form $\int_C P\,dx + Q\,dy$ is path independent if

$$\frac{\partial P}{\partial y} = \frac{\partial Q}{\partial x}.$$

Here $P = y^2 + kxy$, $Q = x^2 + kxy$, so

$$\frac{\partial}{\partial y}(y^2 + kxy) = \frac{\partial}{\partial x}(x^2 + kxy)$$

or $2y + kx = 2x + ky$, whence $k = 2$.

15.3 Evaluate $\int_C (3x + 2xy)\,ds$, where s is a measure of arc length and C is (a) the line $y = x$ from $(0,0)$ to $(1,1)$, (b) $y = x^2$ from $(0,0)$ to $(1,1)$.

In general, $ds^2 = dx^2 + dy^2 = \sqrt{1 + (dy/dx)^2}\,dx$.

(a) $\displaystyle\int_C = \int_{x=0}^{x=1} (3x + 2x^2)\sqrt{1 + (1)^2}\,dx$ (since $y = x$) $= \dfrac{13\sqrt{2}}{6}$.

(b) $\displaystyle\int_C = \int_0^1 (3x + 2x^3)\sqrt{1 + (2x)^2}\,dx = \int_0^1 (3x + 2x^3)\sqrt{1 + 4x^2}\,dx.$

Let $x^2 = u$. Then the integral is

$$\frac{3}{2}\int_0^1 \sqrt{1+4u}\,du + \int_0^1 u\sqrt{1+4u}\,du = \tfrac{3}{2}[\tfrac{2}{3} \cdot \tfrac{1}{4}(1+4u)^{3/2}]_0^1$$

$$+ [\tfrac{2}{3} \cdot \tfrac{1}{4}u(1+4u)^{3/2}]_0^1 - \tfrac{1}{6}\int_0^1 (1+4u)^{3/2}\,du,$$

where the second integral has been evaluated by parts. Carrying out the elementary integration $\int_0^1 (1+4u)^{3/2}\,du$ and inserting the limits

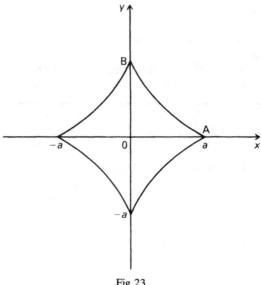

Fig. 23

we have finally

$$\int_C = \frac{5\sqrt{5}}{3} - \frac{7}{30}.$$

15.4 Evaluate $\oint_C (x\,dy - y\,dx)$, where C is the curve $x = a\cos^3 t$, $y = a\sin^3 t$ (see Fig. 23).

At A, $t = 0$; at B, $t = \pi/2$. Hence

$$\int_A^B (x\,dy - y\,dx) = \int_{t=0}^{\pi/2} [a\cos^3 t\, 3a\sin^2 t\cos t\, dt$$

$$- a\sin^3 t\,(-3a\cos^2 t\sin t)\,dt]$$

$$= 3a^2 \int_0^{\pi/2} (\cos^4 t\sin^2 t + \sin^4 t\cos^2 t)\,dt.$$

The integral around total path is $4\int_A^B$ since the integrand is an even function. Simplifying the integrand we have therefore

$$\oint_C (x\,dy - y\,dx) = 12a^2 \int_0^{\pi/2} (\cos^2 t - \cos^4 t)\,dt$$

$$= 12a^2 \left(\frac{1}{2}\cdot\frac{\pi}{2} - \frac{3}{4}\cdot\frac{1}{2}\cdot\frac{\pi}{2}\right) = \frac{3\pi a^2}{4}.$$

100

15.5 Given that

$$u = yF(xy) - \frac{y}{x^2} G\left(\frac{y}{x}\right), \qquad v = xF(xy) + \frac{1}{x} G\left(\frac{y}{x}\right),$$

show that the line integral $\int_L u \, dx + v \, dy$ depends only on the end points of L, which is solely in the region $x > 0$. Show that

$$\phi(x, y) = \int_{(1,1)}^{(x,y)} u \, dx + v \, dy = \int_1^{xy} F(t) \, dt + \int_1^{y/x} G(t) \, dt,$$

and evaluate the change in $\phi(x, y)$ between the points (a, a) and $(a/b, ab)$ in the case $G(t) = 1/t$. (L.U.)

Path independent if $\dfrac{\partial u}{\partial y} = \dfrac{\partial v}{\partial x}$.

$$\frac{\partial v}{\partial x} = xyF'(xy) + F(xy) - \frac{1}{x^2} G\left(\frac{y}{x}\right) - \frac{y}{x^3} G'\left(\frac{y}{x}\right),$$

$$\frac{\partial u}{\partial y} = xyF'(xy) + F(xy) - \frac{1}{x^2} G\left(\frac{y}{x}\right) - \frac{y}{x^3} G'\left(\frac{y}{x}\right),$$

($F'(xy)$ denotes derivative with respect to the argument xy etc.). Hence path independent.

$$\phi(x, y) = \int_{(1,1)}^{(x,y)} \left(yF(xy) - \frac{y}{x^2} G\left(\frac{y}{x}\right)\right) dx + \int_{(1,1)}^{(x,y)} \left(xF(xy) + \frac{1}{x} G\left(\frac{y}{x}\right)\right) dy$$

$$= \int_{(1,1)}^{(x,y)} \left[(y \, dx + x \, dy)F(xy) + \left(-\frac{y}{x^2} dx + \frac{1}{x} dy\right) G\left(\frac{y}{x}\right) \right]$$

$$= \int_{(1,1)}^{(x,y)} F(xy) \, d(xy) + \int_{(1,1)}^{(x,y)} G\left(\frac{y}{x}\right) d\left(\frac{y}{x}\right)$$

$$= \int_{t=1}^{t=xy} F(t) \, dt + \int_{t=1}^{t=y/x} G(t) \, dt.$$

$$\phi(a, a) = \int_{t=1}^{a^2} F(t) \, dt + \int_{t=1}^{1} G(t) \, dt,$$

$$\phi\left(\frac{a}{b}, ab\right) = \int_{t=1}^{a^2} F(t) \, dt + \int_{t=1}^{b^2} G(t) \, dt$$

The change in ϕ is

$$\phi\left(\frac{a}{b}, ab\right) - \phi(a, a) = \int_1^{b^2} G(t) \, dt = \int_1^{b^2} \frac{1}{t} \, dt = 2 \log_e b.$$

15.6 A vector field is given by

$$\mathbf{F} = (2xy + y)\mathbf{i} + 4x^2 y\mathbf{j}.$$

Evaluate the integral $\int_C \mathbf{F} \cdot \mathbf{dr}$ from the point $(0, 0)$ to $(1, 1)$, where the path C is along (a) $y = x$, (b) $y = x^2$. Determine whether the integral is path independent or not.

$$\mathbf{dr} = \mathbf{i}\,dx + \mathbf{j}\,dy, \qquad \int_C \mathbf{F} \cdot \mathbf{dr} = \int_C (2xy + y)\,dx + 4x^2 y\,dy.$$

(a) $\displaystyle \int_0^1 (2x^2 + x)\,dx + \int_0^1 4x^3\,dx = \tfrac{2}{3} + \tfrac{1}{2} + 1 = \tfrac{13}{6}.$

(b) $\displaystyle \int_0^1 (2x^3 + x^2)\,dx + \int_0^1 4y^2\,dy = \tfrac{1}{2} + \tfrac{1}{3} + \tfrac{4}{3} = \tfrac{13}{6}.$

For path independence we require $\dfrac{\partial}{\partial y}(2xy + y) = \dfrac{\partial}{\partial x}(4x^2 y)$, which is not so. Hence the integral is not path independent (even though the results along the two given paths are equal).

15.7 Evaluate $\iint xy\,dx\,dy$ taken over the area enclosed by the coordinate axes and the line $\dfrac{x}{a} + \dfrac{y}{b} = 1$ (see Fig. 24).

Expressed as a repeated integral

$$\int_{x=0}^{x=a} dx \int_{y=0}^{y=(1-x/a)b} xy\,dy = \tfrac{1}{2}\int_0^a xb^2 \left(1 - \frac{x}{a}\right)^2 dx$$

$$= \frac{b^2}{2}\left[\frac{x^2}{2} - \frac{2}{3}\frac{x^3}{a} + \frac{x^4}{4a}\right]_0^a = \frac{a^2 b^2}{24}.$$

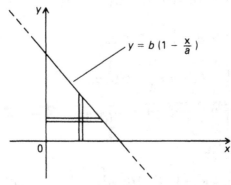

$$y = b\left(1 - \frac{x}{a}\right)$$

Fig. 24

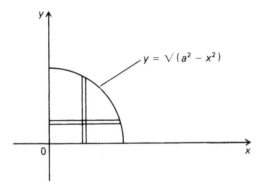

Fig. 25

15.8 Find the value of $\iint (a^2 - x^2)\,dx\,dy$ taken over the area enclosed by the first quadrant of the circle $x^2 + y^2 = a^2$ (see Fig. 25).

Integrating first with respect to y (up the vertical strip) and then with respect to x (across the horizontal strip) we have

$$I = \iint (a^2 - x^2)\,dx\,dy = \int_{x=0}^{a} dx \int_{y=0}^{\sqrt{a^2-x^2}} (a^2 - x^2)\,dy$$

$$= \int_{x=0}^{a} (a^2 - x^2)^{3/2}\,dx.$$

Let $x = a \sin \theta$. Then integral is

$$I = \int_{\theta=0}^{\pi/2} (a^2 - a^2 \sin^2 \theta)^{3/2} a \cos \theta\,d\theta = a^4 \int_0^{\pi/2} \cos^4 \theta\,d\theta = \frac{3\pi a^4}{16}.$$

15.9 By inverting the order of integration evaluate

(a) $\displaystyle\int_0^1 dy \int_0^{\cos^{-1} y} \sec x\,dx,$ where $0 \leqslant \cos^{-1} y \leqslant \pi/2.$

(b) $\displaystyle\int_0^1 dy \int_y^1 \frac{y}{x} e^x\,dx.$

(a) In Fig. 26, instead of integrating along the horizontal strip from $x = 0$ to $x = \cos^{-1} y$, integrate first up the vertical strip from $y = 0$ to $y = \cos x$. Then

$$\int_0^1 dy \int_0^{\cos^{-1} y} \sec x\,dx = \int_{x=0}^{\pi/2} dx \int_{y=0}^{\cos x} \sec x\,dy$$

$$= \int_0^{\pi/2} \sec x \cos x\,dx = \pi/2.$$

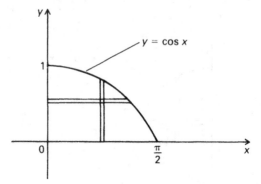

Fig. 26

(b) In Fig. 27, integrate first vertically from $y = 0$ to $y = x$, and then with respect to x from $x = 0$ to $x = 1$. Hence

$$\int_0^1 dy \int_y^1 \frac{y}{x} e^x \, dx = \int_{x=0}^1 dx \int_{y=0}^x \frac{y \, e^x}{x} \, dy = \int_{x=0}^1 \frac{e^x}{x} \left(\frac{x^2}{2} \right) dx$$

$$= \frac{1}{2} \int_0^1 x e^x \, dx = \tfrac{1}{2} [x e^x - e^x]_0^1 = \tfrac{1}{2}.$$

15.10   Evaluate the integral

$$\int_{-\infty}^{\infty} \int_{-\infty}^{\infty} \frac{x^2 \, dx \, dy}{(1 + \sqrt{x^2 + y^2})^5}$$

by transforming the Cartesian variables x, y to polar coordinates r, θ.

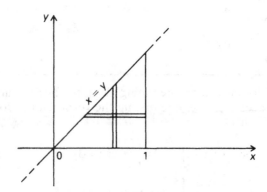

Fig. 27

Basic result

If $x = x(u, v)$ and $y = y(u, v)$, where u and v are new variables, then

$$\iint_{R_{xy}} f(x, y)\, dx\, dy = \iint_{R_{uv}} f[x(u, v), y(u, v)] \left| J\!\left(\frac{x, y}{u, v}\right) \right| du\, dv,$$

where $\left| J\!\left(\dfrac{x, y}{u, v}\right) \right|$ is the modulus of the Jacobian determinant (see 12.4–12.6), and R_{uv} is the region in the uv-plane corresponding to R_{xy} in the xy-plane. The region of integration for the integral given is the whole of the xy-plane. In terms of the r, θ coordinates, this region corresponds to $0 \leqslant r < \infty$, $0 \leqslant \theta \leqslant 2\pi$. Hence using $x = r \cos \theta$, $y = r \sin \theta$, we have

$$J\!\left(\frac{x, y}{r, \theta}\right) = \frac{\partial(x, y)}{\partial(r, \theta)} = \begin{vmatrix} \dfrac{\partial x}{\partial r} & \dfrac{\partial x}{\partial \theta} \\[2mm] \dfrac{\partial y}{\partial r} & \dfrac{\partial y}{\partial \theta} \end{vmatrix} = \begin{vmatrix} \cos \theta & -r \sin \theta \\ \sin \theta & r \cos \theta \end{vmatrix} = r.$$

The basic result stated above now gives

$$I = \int_{-\infty}^{\infty} \int_{-\infty}^{\infty} \frac{x^2\, dx\, dy}{(1 + \sqrt{x^2 + y^2})^5} = \int_{r=0}^{\infty} \int_{\theta=0}^{2\pi} \frac{r^2 \cos^2 \theta}{(1 + r)^5}\, r\, d\theta\, dr$$

$$= \int_0^{2\pi} \cos^2 \theta\, d\theta \int_{r=0}^{\infty} \frac{r^3\, dr}{(1 + r)^5}.$$

By putting $1 + r = u$ and integrating, we find $I = \pi/4$.

15.11 Show by means of a diagram the region over which the integral

$$\int_0^2 dy \int_y^{4-y} \left(1 + \frac{y}{x}\right)^2 e^{(x^2 - y^2)/x}\, dx$$

is taken. Using the transformation of variables

$$u = x + y, \quad v = y/x,$$

or otherwise, evaluate the integral. (L.U.)

Now $u = x + y$, $v = y/x$ implies $x = u/(1 + v)$, $y = uv/(1 + v)$. Hence

$$J = \begin{vmatrix} \dfrac{1}{1 + v} & \dfrac{v}{1 + v} \\[3mm] -\dfrac{u}{(1 + v)^2} & \dfrac{u}{(1 + v)^2} \end{vmatrix} = \frac{u}{(1 + v)^2}.$$

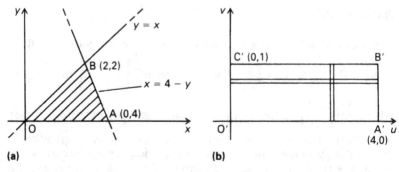

Fig. 28

From Fig. 28(a) and (b) we see that AB ($x = 4 - y$) becomes A'B' ($u = 4$), OB ($x = y$) becomes B'C' ($v = 1$), OA ($y = 0$) becomes O'A' ($v = 0$), and the point O becomes the line O'C' ($u = 0$). The integration is now to be carried out over the rectangle shown in Fig. 28(b). Hence the integral becomes

$$\int_{u=0}^{4} \int_{v=0}^{1} u e^{u(1-v)} \, du \, dv = \int_{u=0}^{4} -[e^{u(1-v)}]^{1}_{v=0} \, du$$

$$= \int_{0}^{4} (e^{u} - 1) \, du = e^{4} - 5.$$

15.12 (a) Use Green's theorem in the plane to evaluate

$$\oint_{C} \{(x^{2} - 2x \cos y) \, dx + (x^{2} \sin y + 3 \sin x + y^{2}) \, dy\},$$

where C is the boundary of the rectangle $0 \leqslant x \leqslant \pi/2$, $0 \leqslant y \leqslant 1$.
(b) Use Green's theorem to evaluate the integral

$$\oint_{C} \frac{(2 - y) \, dx + x \, dy}{x^{2} + (2 - y)^{2}},$$

where C is the circle of unit radius centred on the origin. Calculate the integral directly for a single circuit of the circle of radius a with centre $(0, 2)$.

Basic result

Green's theorem in the plane states that

$$\oint_{C} P \, dx + Q \, dy = \iint_{R} \left(\frac{\partial Q}{\partial x} - \frac{\partial P}{\partial y} \right) dx \, dy,$$

where P and Q are given functions of x and y, and C is the boundary of R.

(a) $P = x^2 - 2x \cos y$, $\quad Q = x^2 \sin y + 3 \sin x + y^2$.

$$\frac{\partial P}{\partial y} = 2x \sin y, \qquad \frac{\partial Q}{\partial x} = 2x \sin y + 3 \cos x.$$

Hence

$$\frac{\partial Q}{\partial x} - \frac{\partial P}{\partial y} = 3 \cos x.$$

Therefore

$$\oint_C = \iint_R 3 \cos x \, dx \, dy = 3 \int_0^{\pi/2} \cos x \, dx \int_0^1 dy = 3.$$

(b) $P = \dfrac{2-y}{x^2 + (2-y)^2}$, $\quad Q = \dfrac{x}{x^2 + (2-y)^2}$.

$$\frac{\partial P}{\partial y} = \frac{-x^2 + (2-y)^2}{[x^2 + (2-y)^2]^2}, \qquad \frac{\partial Q}{\partial x} = \frac{1}{x^2 + (2-y)^2} - \frac{2x^2}{[x^2 + (2-y)^2]^2},$$

so $\partial Q/\partial x = \partial P/\partial y$ at all points, *except at* $(0, 2)$ where P, Q are not defined. Hence if C does *not* enclose the point $(0, 2)$, then the integral is path independent and accordingly $\oint_C = 0$.

If C encloses $(0, 2)$ the conditions of the theorem are not satisfied (i.e. P, Q and their derivatives must exist). Hence we must evaluate the integral directly. Taking the new origin as $(0, 2)$ and writing $x = a \cos \theta$, $\quad y - 2 = a \sin \theta$ $\quad$ we $\quad$ have $\quad$ $dx = -a \sin \theta \, d\theta$, $\quad dy = a \cos \theta \, d\theta$, and hence

$$\oint_C = \int_0^{2\pi} (\sin^2 \theta + \cos^2 \theta) \, d\theta = 2\pi.$$

15.13 Given that

$$\int_{-\infty}^{\infty} f(x) \, dx = 1, \qquad \int_{-\infty}^{\infty} g(x) \, dx = 1,$$

where f and g are arbitrary functions, the convolution integral of f and g is defined as

$$f * g = \int_{-\infty}^{\infty} f(x - x')g(x') \, dx'.$$

Show that

$$\int_{-\infty}^{\infty} f * g \, dx = 1.$$

Consider

$$\int_{-\infty}^{\infty} f*g\,dx = \int_{-\infty}^{\infty}\int_{-\infty}^{\infty} f(x-x')g(x')\,dx'\,dx.$$

Transform to new variables u, v, where $u = x - x'$, $v = x'$. Then

$$J = \frac{\partial(x, x')}{\partial(u, v)} = \begin{vmatrix} 1 & 1 \\ 0 & 1 \end{vmatrix} = 1.$$

Hence

$$\int_{-\infty}^{\infty} f*g\,dx = \int_{u=-\infty}^{\infty}\int_{v=-\infty}^{\infty} f(u)g(v)\,du\,dv$$

$$= \int_{-\infty}^{\infty} f(u)\,du \int_{-\infty}^{\infty} g(v)\,dv = 1.$$

(N.B. From the definition, it follows that $g*f = f*g$).

15.14   Consider the integral

$$I = \frac{1}{2\pi\sqrt{1-\rho^2}} \int_{-\infty}^{h}\int_{-\infty}^{k} \exp\left\{-\frac{1}{2}\left(\frac{x^2 - 2\rho xy + y^2}{1-\rho^2}\right)\right\} dy\,dx$$

(ρ = constant). By means of the transformations $s = x$, $t = (y-\rho x)/\sqrt{1-\rho^2}$, obtain the form

$$I = \int_{-\infty}^{h} \frac{1}{\sqrt{2\pi}} e^{-s^2/2}\left\{\int_{-\infty}^{(k-\rho s)/\sqrt{1-\rho^2}} \frac{e^{-t^2/2}}{\sqrt{2\pi}} dt\right\} ds.$$

Using the result in 15.10, $ds\,dt = |J|\,dx\,dy$, where

$$J = \begin{vmatrix} \dfrac{\partial s}{\partial x} & \dfrac{\partial s}{\partial y} \\ \dfrac{\partial t}{\partial x} & \dfrac{\partial t}{\partial y} \end{vmatrix} = \begin{vmatrix} 1 & 0 \\ \dfrac{-\rho}{\sqrt{1-\rho^2}} & \dfrac{1}{\sqrt{1-\rho^2}} \end{vmatrix} = \frac{1}{\sqrt{1-\rho^2}}.$$

Hence

$$dx\,dy = \sqrt{1-\rho^2}\,ds\,dt.$$

Now inserting for x and y in terms of s and t we have

$$I = \frac{1}{2\pi} \int_{s(=x)=-\infty}^{h}\int_{t=-\infty}^{(k-\rho s)/\sqrt{1-\rho^2}}$$

$$\exp\left\{-\frac{s^2 - 2\rho s(t\sqrt{1-\rho^2}+\rho s) + t^2(1-\rho^2) + 2t\rho s\sqrt{1-\rho^2} + \rho^2 s^2}{2(1-\rho^2)}\right\} ds\,dt$$

(using $x = s$)

$$= \int_{-\infty}^{h} \frac{1}{\sqrt{2\pi}} e^{-s^2/2} \left\{ \int_{-\infty}^{(k-\rho s)/\sqrt{1-\rho^2}} \frac{1}{\sqrt{2\pi}} e^{-t^2/2} \, dt \right\} ds.$$

The upper limit for the t-integral is obtained from the fact that

$$t = \frac{y - \rho x}{\sqrt{1-\rho^2}} = \frac{y - \rho s}{\sqrt{1-\rho^2}}$$

(since $x = s$) and y goes from $-\alpha$ to k.

16. ORDINARY DIFFERENTIAL EQUATIONS

[Formation of differential equations. Types of equations: variable separable, homogeneous, general linear first-order, exact, simple non-linear, second-order constant coefficient, Euler form, simultaneous, matrix form. Picard iterative method, Leibnitz–Maclaurin series, second-order variable coefficient, Frobenius series, normal form, Riccati's equation, Pinney's equation]

16.1. Obtain differential equations of the first-order (independent of arbitrary constants) of which the following functions are the general solutions:

(a) $y = A \sin x$, (b) $y = (x + A) \sin x$,

where A is an arbitrary constant.

(a) Differentiating we have $\dfrac{dy}{dx} = A \cos x$, and eliminating A gives the equation

$$\frac{dy}{dx} = y \cot x.$$

(b) Likewise $\dfrac{dy}{dx} = (x + A) \cos x + \sin x$. But $y \operatorname{cosec} x = x + A$. Hence

$$\frac{dy}{dx} = y \cot x + \sin x.$$

16.2 What is wrong with forming the following differential equations:

(a) If $y = A \log_e \sin x$, where A is an arbitrary constant, then

$$\frac{dy}{dx} = y \frac{\cot x}{\log_e \sin x} \quad \text{for } all \text{ real } x.$$

(b) If $y = A \log_e \{\log_e \sin x\}$, where A is an arbitrary constant and x is real, then

$$\frac{dy}{dx} = \frac{y \cot x}{\log_e \sin x} \frac{1}{\log_e \{\log_e \sin x\}}.$$

(a) The function $\log_e \sin x$ exists only for *some* values of x. For example, if $0 < x < \pi$, then $\sin x$ is positive and its logarithm exists. However, if $\pi < x < 2\pi$ then $\sin x$ is negative and its logarithm does not exist in the real number system. Hence the differential coefficient does not exist for *all* real x.

(b) The function $\log_e \{\log_e \sin x\}$ does not exist for real values of x since $\log_e \sin x$ is either negative (for example, for $0 < x < \pi$) or does not exist (for example, if $\pi < x < 2\pi$). Hence its logarithm does not exist for any real values of x and the differential coefficient is meaningless.

16.3 Obtain differential equations of the second-order (independent of arbitrary constants) of which the following functions are the general solutions:

(a) $y = Ae^{-x} + Bx$, (b) $y = A \sinh(x + B)$,

where A and B are arbitrary constants.

(a) $\dfrac{dy}{dx} = -Ae^{-x} + B, \dfrac{d^2y}{dx^2} = Ae^{-x}.$

Hence $(1 + x)\dfrac{d^2y}{dx^2} + x\dfrac{dy}{dx} - y = 0.$

(b) $\dfrac{dy}{dx} = A \cosh(x + B), \dfrac{d^2y}{dx^2} = A \sinh(x + B).$

Hence $\dfrac{d^2y}{dx^2} = y.$

16.4 Solve the following differential equations:

(a) $\dfrac{dy}{dx} = \dfrac{xe^y}{x^2+1}$ with $y(0) = 0$,

(b) $2\dfrac{dy}{dx} = \dfrac{y}{x} + \left(\dfrac{y}{x}\right)^2$,

(c) $\dfrac{dy}{dx} = \dfrac{x+y-1}{x-y-2}$,

(d) $(1-x^2)\dfrac{dy}{dx} + xy = x$ with $y(0) = 2$.

(a) Variable separable type.

$$\int e^{-y}\, dy = \int \frac{x\, dx}{x^2+1} = \tfrac{1}{2}\int \frac{d(x^2)}{x^2+1} = \tfrac{1}{2}\log_e (1+x^2) + C.$$

Hence $e^{-y} = -C - \tfrac{1}{2}\log_e (1+x^2)$. Initial condition $y(0) = 0$ gives $C = -1$. Hence solution is

$$e^{-y} = 1 - \tfrac{1}{2}\log_e (1+x^2).$$

(b) Homogeneous type. Put $y = vx$, then $dy/dx = v + x\, dv/dx$. The equation now becomes $2(v + x\, dv/dx) = v + v^2$, which is of variable separable type. Hence

$$\int \frac{dv}{v^2 - v} = \int \frac{dx}{2x}.$$

Writing the left-hand side as $\displaystyle\int \frac{dv}{(v-\tfrac{1}{2})^2 - \tfrac{1}{4}}$ and integrating we find

$$-\log_e \left(\frac{v}{v-1}\right) = \tfrac{1}{2}\log_e x + \log_e C,$$

where C is an arbitrary constant. The solution is therefore

$$Cx^{1/2}\left(\frac{v}{v-1}\right) = 1, \quad \text{or} \quad Cx^{1/2}y = y - x.$$

(c) Reducible to homogeneous form. Let $X + h = x$, $Y + k = y$. Then $\dfrac{dy}{dx} = \dfrac{dY}{dX}$. $x + y - 1 = X + h + Y + k - 1$, $x - y - 2 = X - Y + h - k - 2$. Now choose $h + k = 1$, $h - k = 2$ so that $h = \tfrac{3}{2}$, $k = -\tfrac{1}{2}$. Hence the equation is

$$\frac{dY}{dX} = \frac{X+Y}{X-Y},$$

which is of homogeneous form. Let $Y = vX$ (as in (b)), whence (since $dY/dX = v + X \, dv/dX$)

$$X\frac{dv}{dX} = \frac{1 + v^2}{1 - v}.$$

Integrating we get

$$\log_e X = \tan^{-1} v - \tfrac{1}{2}\log_e(1 + v^2) + C,$$

where C is an arbitrary constant. Hence finally substituting back for x and y we have

$$\log_e(x - \tfrac{3}{2}) = \tan^{-1}\left(\frac{y + \tfrac{1}{2}}{x - \tfrac{3}{2}}\right) - \tfrac{1}{2}\log_e\left[1 + \frac{(y + \tfrac{1}{2})^2}{(x - \tfrac{3}{2})^2}\right] + C.$$

(d) General linear first-order form $dy/dx + P(x)y = Q(x)$.

Basic result

The integrating factor is $R(x) = e^{\int P \, dx}$ and the solution is $Ry = \int RQ \, dx + C$, where C is an arbitrary constant of integration. Here $P = \dfrac{x}{1 - x^2}$, $Q = \dfrac{x}{1 - x^2}$, so $R = e^{\int x/(1 - x^2) \, dx} = e^{-\frac{1}{2}\log_e(1 - x^2)} = \dfrac{1}{\sqrt{1 - x^2}}$.
The solution is therefore

$$\frac{y}{\sqrt{1 - x^2}} = \int \frac{x \, dx}{(1 - x^2)^{3/2}} + C = \frac{1}{\sqrt{1 - x^2}} + C.$$

Since $y = 2$ when $x = 0$, $C = 1$, and therefore $y = 1 + \sqrt{1 - x^2}$.

16.5 Find the value of k which makes the differential equation $(k + xy)y \, dx + [x + y(1 + x^2)] \, dy = 0$ exact. Obtain y in terms of x for this value of k, given that $y = 1$ when $x = 0$.

Write $(k + xy)y = P$, $x + y(1 + x^2) = Q$. Then condition for exactness is $\partial P/\partial y = \partial Q/\partial x$. Hence $k + 2xy = 1 + 2xy$, giving $k = 1$. For the equation to have the form $d\phi = 0$ we require $\partial \phi/\partial x = P$, $\partial \phi/\partial y = Q$. Hence

$$\phi = \int P \, dx = \int (1 + xy)y \, dx = xy + \tfrac{1}{2}x^2y^2 + u(y),$$

where $u(y)$ is an arbitrary function. Similarly

$$\phi = \int Q \, dy = \int [x + y(1 + x^2)] \, dy = xy + (1 + x^2)\frac{y^2}{2} + v(x),$$

where $v(x)$ is an arbitrary function. The two forms for ϕ are identical if we choose $u = \frac{1}{2}y^2$, $v = 0$, whence

$\phi = xy + \frac{1}{2}y^2(1 + x^2) = C$ (constant)

is the required solution. Now $y = 1$ at $x = 0$ so $C = \frac{1}{2}$. Hence finally

$$xy + \frac{1}{2}y^2(1 + x^2) = \frac{1}{2} \quad \text{or} \quad y = \frac{\sqrt{1 + 2x^2} - x}{1 + x^2}.$$

16.6 Solve the following non-linear equations by transforming them to a suitable linear form:

(a) $\dfrac{d^2y}{dx^2} + 2\left(\dfrac{dy}{dx}\right)^2 = y^2$ given $\dfrac{dy}{dx} = \frac{1}{4}$ when $y = 0$, and $y = \frac{1}{4}$ at $x = 0$,

(b) $x\dfrac{d^2y}{dx^2} = (1 + y)\dfrac{dy}{dx}$ given $y = 0$, $\dfrac{dy}{dx} = 0$ when $x = 0$,

(c) $(y + y') \log_e (y + y') + y = 0$, $(y' = dy/dx)$.

(a) Put $dy/dx = p$. Then $d^2y/dx^2 = p\,(dp/dy)$. Hence the equation becomes

$$p\frac{dp}{dy} + 2p^2 = y^2 \quad \text{or} \quad \frac{1}{2}\frac{d}{dy}(p^2) + 2p^2 = y^2.$$

Writing $p^2 = u$, we have the linear equation for u in terms of y:

$$\frac{du}{dy} + 4u = 2y^2.$$

In terms of the standard notation (see 16.4) $P = 4$, $Q = 2y^2$, and the integrating factor $R = e^{\int P\,dy} = e^{4y}$. Hence the solution is

$$ue^{4y} = 2\int y^2 e^{4y}\,dy + C$$

whence $u = Ce^{-4y} + \left(\dfrac{y^2}{2} - \dfrac{y}{4} + \frac{1}{16}\right)$, where C is an arbitrary constant. Now $u = p^2 = \frac{1}{16}$ when $y = 0$. Hence $C = 0$. Therefore

$$p^2 = \frac{y^2}{2} - \frac{y}{4} + \frac{1}{16}.$$

Integrating (since $p = dy/dx$) we have

$$\sqrt{2}\int \frac{dy}{\sqrt{(y - \frac{1}{4})^2 + \frac{1}{16}}} = \sqrt{2}\sinh^{-1}\left[\frac{y - \frac{1}{4}}{\frac{1}{4}}\right] = x + A$$

(using the standard integral $\int (1/\sqrt{x^2 + a^2})\,dx = \sinh^{-1}(x/a)$), where

113

A is an arbitrary constant. Consequently

$$4y - 1 = \sinh\left(\frac{x+A}{\sqrt{2}}\right).$$

Now $y = \frac{1}{4}$ at $x = 0$ so $A = 0$. Hence

$$y = \tfrac{1}{4}(1 + \sinh(x/\sqrt{2})).$$

(b) $x\dfrac{d^2y}{dx^2} = (1+y)\dfrac{dy}{dx}.$

Now

$$\frac{d}{dx}\left(x\frac{dy}{dx}\right) = x\frac{d^2y}{dx^2} + \frac{dy}{dx}.$$

Hence

$$\frac{d}{dx}\left(x\frac{dy}{dx}\right) - \frac{dy}{dx} = (1+y)\frac{dy}{dx}$$

or

$$\frac{d}{dx}\left(x\frac{dy}{dx}\right) = 2\frac{dy}{dx} + y\frac{dy}{dx}.$$

Integrating:

$$x\frac{dy}{dx} = 2y + \tfrac{1}{2}y^2 + A.$$

But

$y = 0$, $dy/dx = 0$ at $x = 0$. Hence $A = 0$. Therefore

$$x\frac{dy}{dx} = 2y + \tfrac{1}{2}y^2.$$

Therefore

$$2\frac{dy}{dx} = \frac{y^2 + 4y}{x}$$

or

$$2\int \frac{dy}{(y+2)^2 - 4} = \int \frac{dx}{x}.$$

Integrating gives $\log_e\left(\dfrac{y}{y+4}\right) = 2\log_e x + C$, where C is an arbitrary constant. Hence finally

$$y = \frac{4Cx^2}{1 - Cx^2}.$$

(c) Differentiating we find

$$(y' + y'') \log_e (y + y') + (y' + y'') + y' = 0.$$

Inserting for $\log_e (y + y')$ from the given equation, we have after some simplification

$$y'(y'' + 2y' + y) = 0.$$

Hence $y' = 0$ or $y'' + 2y' + y = 0$. Now the original equation is not satisfied by $y = 0$ since this implies $y' = 0$ and $\log_e (y + y')$ is not defined. The equation therefore reduces to $y'' + 2y' + y = 0$, which by D-operator methods (see 16.7) has the solution $y = (Ax + B)e^{-x}$, where A and B are arbitrary constants. Now $y = e^{-x}$ does not satisfy the equation since again $\log_e (y + y')$ is not defined. Hence B must be put equal to zero. The solution is therefore $y = Axe^{-x}$, where A must be taken equal to unity to satisfy the equation.

16.7 Solve, using the D-operator method, the differential equation

$$\frac{d^2y}{dx^2} + \frac{dy}{dx} - 6y = \sin x + xe^{2x}.$$

Basic results

Given a constant coefficient equation of the type

$$a_0 \frac{d^n y}{dx^n} + a_1 \frac{d^{n-1} y}{dx^{n-1}} + \cdots + a_{n-1} \frac{dy}{dx} + a_n y = f(x),$$

where $a_0, a_1, \ldots, a_n$ are constants, we write this in the form

$$F(D)y = f(x),$$

where $D \equiv d/dx$ and $F(D) = a_0 D^n + a_1 D^{n-1} + \cdots + a_{n-1} D + a_n$.

The solution is the sum of the complementary function (C.F.) which is the solution of the homogeneous equation $F(D)y = 0$, and the particular integral (P.I.) which is any one solution of $F(D)y = f(x)$. To obtain the C.F. we let $y = e^{mx}$ and solve for the values of m. To find a particular integral when $f(x)$ has simple forms such as e^{kx}, $\sin kx$, $\cos kx$, x^n and their respective products we use the three basic theorems of the D-operator method:

1. $F(D)e^{kx} = e^{kx}F(k)$, where k is a constant;

2. $F(D)[e^{kx}V(x)] = e^{kx}F(D+k)V(x)$, where $V(x)$ is an arbitrary function;

3. $F(D^2)\begin{Bmatrix} \sin kx \\ \cos kx \end{Bmatrix} = F(-k^2)\begin{Bmatrix} \sin kx \\ \cos kx \end{Bmatrix}$.

In this particular case the C.F. is the solution of the equation

$$\frac{d^2y}{dx^2} + \frac{dy}{dx} - 6y = 0.$$

Letting $y = e^{mx}$ then $m^2 + m - 6 = 0$ (auxiliary equation). Hence $m = 2, -3$. Therefore C.F. is $y = Ae^{2x} + Be^{-3x}$, where A and B are arbitrary constants. Now $F(D) = D^2 + D - 6$ so the P.I. is obtained by evaluating

$$y = \frac{1}{D^2 + D - 6}\sin x + \frac{1}{D^2 + D - 6}xe^{2x}$$

$$= (\text{using Th.3})\frac{1}{-1 + D - 6}\sin x$$

$$+ (\text{using Th. 2})e^{2x}\frac{1}{(D+2)^2 + (D+2) - 6}x$$

$$= \frac{1}{D-7}\sin x + e^{2x}\left(\frac{1}{D^2 + 5D}\right)x$$

$$= \frac{D+7}{D^2 - 49}\sin x + \frac{e^{2x}}{5}\left(\frac{1}{D} - \frac{1}{D+5}\right)x$$

$$= -\tfrac{1}{50}(\cos x + 7\sin x) + \frac{e^{2x}}{5}\left(\frac{x^2}{2} - \frac{x}{5} + \frac{1}{25}\right).$$

16.8 Obtain the general solution of

$$x^2\frac{d^2y}{dx^2} - x\frac{dy}{dx} + 4y = \cos(\log_e x).$$

This equation is an example of the Euler form where the *powers* of x multiplying the differential coefficients are equal to the *order* of the differential coefficients. Let $x = e^t$. Then

$$\frac{dy}{dx} = \frac{dy}{dt}\frac{dt}{dx} = \frac{1}{x}\frac{dy}{dt} \quad \text{so} \quad x\frac{dy}{dx} = \frac{dy}{dt}.$$

Similarly

$$x^2\frac{d^2y}{dx^2} = \frac{d^2y}{dt^2} - \frac{dy}{dt}.$$

Therefore equation becomes

$$\frac{d^2y}{dt^2} - 2\frac{dy}{dt} + 4y = \cos t,$$

which is now a constant coefficient equation for y in terms of t. Solving this as in 16.7, we find that the auxiliary equation is

$$m^2 - 2m + 4 = 0,$$

whence

$$m = \frac{2 \pm \sqrt{4 - 16}}{2} = (1 \pm i\sqrt{3}).$$

Hence C.F. is

$$y = e^t(A \cos \sqrt{3}t + B \sin \sqrt{3}t)$$

(A and B are arbitrary constants). Now the P.I. is

$$y = \frac{1}{D^2 - 2D + 4} \cos t = \text{Re} \left\{ \frac{1}{D^2 - 2D + 4} e^{it} \right\} = \tfrac{3}{13} \cos t - \tfrac{2}{13} \sin t$$

$$\text{(using Th. 1. of 16.7),}$$

where Re denotes the real part of the complex term following it (e.g. Re $e^{it} = \cos t$). Hence solution is (using $x = e^t$ or $t = \log_e x$)

$$y = x[A \cos (\sqrt{3} \log_e x) + B \sin (\sqrt{3} \log_e x)]$$
$$+ \tfrac{1}{13}(3 \cos \log_e x - 2 \sin \log_e x).$$

16.9 Find n such that $y = x^n$ is a solution of

$$(x^2 - 2x)\frac{d^2y}{dx^2} + 2(1 - x)\frac{dy}{dx} + 2y = 0,$$

and hence find the general solution.

Inserting $y = x^n$ into the equation gives

$$(x^2 - 2x)n(n - 1)x^{n-2} + 2(1 - x)nx^{n-1} + 2x^n = 0$$

or

$$(n^2 - 3n + 2)x^n - (2n^2 - 4n)x^{n-1} = 0,$$

whence $(n - 2)(n - 1) = 0$ and $n(n - 2) = 0$. Hence $n = 2$. To find a second solution write $y = x^2v(x)$, where $v(x)$ is a function to be found. Hence

$$\frac{dy}{dx} = x^2\frac{dv}{dx} + 2xv, \qquad \frac{d^2y}{dx^2} = x^2\frac{d^2v}{dx^2} + 4x\frac{dv}{dx} + 2v.$$

The equation then simplifies to

$$(x^2 - 2x)\frac{d^2v}{dx^2} + 2(x-3)\frac{dv}{dx} = 0$$

or

$$\frac{v''}{v'} = -\frac{2(x-3)}{x(x-2)} = -\left[\frac{3}{x} - \frac{1}{x-2}\right].$$

Integrating

$$\log_e v' = -3\log_e x + \log_e(x-2) + \log_e C \quad \text{or} \quad v' = \frac{C(x-2)}{x^3},$$

where C is an arbitrary constant. Hence $v = C(1/x^2 - 1/x)$, giving the second solution

$$y = x^2 v = C(1-x),$$

and the general solution $y = Ax^2 + B(1-x)$, where A and B are arbitrary constants.

16.10 By writing $y = uv$ and choosing u to make the coefficient of dv/dx vanish, put the equation

$$\frac{d^2y}{dx^2} - \frac{1}{x+1}\frac{dy}{dx} - \frac{x^2+2x+2}{4(x+1)^2}y = 0$$

into normal form. Hence find the solution satisfying $y = 0$ and $dy/dx = 1$ at $x = 0$.

$y = uv$ so

$$\frac{dy}{dx} = u\frac{dv}{dx} + v\frac{du}{dx} \quad \text{and} \quad \frac{d^2y}{dx^2} = u\frac{d^2v}{dx^2} + 2\frac{du}{dx}\frac{dv}{dx} + v\frac{d^2u}{dx^2}.$$

Hence

$$u\frac{d^2v}{dx^2} + \left(2\frac{du}{dx} - \frac{u}{x+1}\right)\frac{dv}{dx} + \left(\frac{d^2u}{dx^2} - \frac{1}{x+1}\frac{du}{dx} - \frac{x^2+2x-2}{4(x+1)^2}u\right)v = 0.$$

Now choose $2\,du/dx - u/(x+1) = 0$, which by integration gives $u = (x+1)^{1/2}$. Hence

$$\frac{d^2v}{dx^2} + \left(-\frac{1}{4(x+1)^2} - \frac{1}{2(x+1)^2} - \frac{x^2+2x-2}{4(x+1)^2}\right)v = 0$$

or

$$\frac{d^2v}{dx^2} - \tfrac{1}{4}v = 0.$$

The solution of this equation is

$$v = A \cosh \frac{x}{2} + B \sinh \frac{x}{2},$$

whence

$$y = (x+1)^{1/2} \left(A \cosh \frac{x}{2} + B \sinh \frac{x}{2} \right).$$

Now $y = 0$, $v = 0$ at $x = 0$ so $A = 0$. Hence

$$y = B(x+1)^{1/2} \sinh \frac{x}{2}.$$

Differentiating and using the given condition $dy/dx = 1$ at $x = 0$ we find $B = 2$. The required solution is therefore

$$y = 2(x+1)^{1/2} \sinh \frac{x}{2}.$$

16.11 Find the general solution of the coupled equations

$$t \frac{dx}{dt} + y = 0, \qquad t \frac{dy}{dt} + x = 0.$$

Differentiating the first equation we have

$$t \frac{d^2 x}{dt^2} + \frac{dx}{dt} + \frac{dy}{dt} = 0,$$

which may be written as

$$t^2 \frac{d^2 x}{dt^2} + t \frac{dx}{dt} - x = 0,$$

making use of the second equation for the term dy/dt. Substituting $t = e^z$ we have

$$t \frac{d}{dt} \equiv \frac{d}{dz} \quad \text{and} \quad t^2 \frac{d^2}{dt^2} \equiv \frac{d^2}{dz^2} - \frac{d}{dz}.$$

Then $d^2 x/dz^2 - x = 0$, whence $x = A e^z + B e^{-z} = At + Bt^{-1}$, and $y = -t \, dx/dt = -At + Bt^{-1}$, where A and B are arbitrary constants.

16.12 The vector $\mathbf{r} = (x, y, z)$ satisfies the vector equation

$$m \frac{d^2 \mathbf{r}}{dt^2} = e\mathbf{E} + \frac{e}{c} \left(\frac{d\mathbf{r}}{dt} \wedge \mathbf{H} \right),$$

where $\mathbf{E} = (0, E, 0)$, $\mathbf{H} = (0, 0, H)$, and e, m, c, E, H are constants. Write the equation in component form and show by solving the equation that

$$x = \frac{cEt}{H} - \frac{mc^2 E}{eH^2} \sin\left(\frac{eH}{mc} t\right),$$

$$y = \frac{mc^2 E}{eH^2} \left\{1 - \cos\left(\frac{eH}{mc} t\right)\right\},$$

$$z = 0,$$

given that $\mathbf{r} = 0$, $\dot{\mathbf{r}} = 0$ at $t = 0$.

The vector product

$$\frac{d\mathbf{r}}{dt} \wedge \mathbf{H} = \begin{vmatrix} \mathbf{i} & \mathbf{j} & \mathbf{k} \\ \dfrac{dx}{dt} & \dfrac{dy}{dt} & \dfrac{dz}{dt} \\ 0 & 0 & H \end{vmatrix} = \mathbf{i} H \frac{dy}{dt} - \mathbf{j} H \frac{dx}{dt}.$$

Hence the equations become in component form

$$m \frac{d^2 x}{dt^2} = \frac{eH}{c} \frac{dy}{dt}, \quad m \frac{d^2 y}{dt^2} = eE - \frac{eH}{c} \frac{dx}{dt}, \quad m \frac{d^2 z}{dt^2} = 0.$$

Integrating the last of these equations we get $z = At + B$, where A and B are arbitrary constants. But $\mathbf{r} = 0$, $\dot{\mathbf{r}} = 0$ at $t = 0$ so $A = B = 0$. Hence $z = 0$. Now

$$\frac{d^2 x}{dt^2} = \frac{eH}{mc} \frac{dy}{dt}, \quad \frac{d^2 y}{dt^2} = \frac{e}{m} E - \frac{eH}{mc} \frac{dx}{dt}.$$

Therefore $dx/dt = (eH/mc)y + C$, where C is an arbitrary constant. But $dx/dt = 0$ when $y = 0$ so $C = 0$. Hence

$$\frac{dx}{dt} = \frac{eH}{mc} y.$$

Inserting this into the equation for $d^2 y/dt^2$, we find

$$\frac{d^2 y}{dt^2} = \frac{eE}{m} - \left(\frac{eH}{mc}\right)^2 y$$

or

$$\left[D^2 + \left(\frac{eH}{mc}\right)^2\right] y = \frac{eE}{m}, \quad \left(D \equiv \frac{d}{dt}\right).$$

This is a constant coefficient equation whose solution is

$$y = \alpha \cos\left(\frac{eHt}{mc}\right) + \beta \sin\left(\frac{eHt}{mc}\right) + \frac{mc^2 E}{eH^2},$$

where α and β are arbitrary constants. But $y = 0$ at $t = 0$ so

$\alpha = -mc^2 E/(eH^2)$. Likewise $\dot{y} = 0$ at $t = 0$, so $\beta = 0$. Hence finally

$$y = \frac{mc^2 E}{eH^2} \left\{ 1 - \cos \left(\frac{eH}{mc} t \right) \right\}$$

and hence

$$x = \frac{cEt}{H} - \frac{mc^2 E}{eH^2} \sin \left(\frac{eH}{mc} t \right).$$

16.13 By putting $dx_1/dt = x_2$, show that the differential equation

$$\frac{d^2 x_1}{dt^2} - 2 \frac{dx_1}{dt} - 3x_1 = 0$$

may be written as $d\mathbf{X}/dt = A\mathbf{X}$, where $\mathbf{X} = \begin{pmatrix} x_1 \\ x_2 \end{pmatrix}$ and A is a matrix. Find A and show that the solution of the differential equation is given by

$$\mathbf{X} = e^{At} \mathbf{X}_0,$$

where $\mathbf{X}_0$ is an arbitrary constant vector, and

$$e^{At} = I + At + \frac{1}{2!} A^2 t^2 + \frac{1}{3!} A^3 t^3 + \cdots$$

By completely solving the differential equation, or otherwise, obtain the matrix e^{At}. (L.U.)

Now

$$\frac{d^2 x_1}{dt^2} = \frac{dx_2}{dt}.$$

Therefore the equation and the relation $dx_1/dt = x_2$ may together be written as

$$\begin{cases} \dfrac{dx_1}{dt} = 0x_1 + x_2, \\[2mm] \dfrac{dx_2}{dt} = 3x_1 + 2x_2, \end{cases}$$

or in matrix form as

$$\frac{d}{dt} \begin{pmatrix} x_1 \\ x_2 \end{pmatrix} = \begin{pmatrix} 0 & 1 \\ 3 & 2 \end{pmatrix} \begin{pmatrix} x_1 \\ x_2 \end{pmatrix}.$$

Consequently

$$A = \begin{pmatrix} 0 & 1 \\ 3 & 2 \end{pmatrix}.$$

If $\mathbf{X} = e^{At}\mathbf{X}_0$,

$$\frac{d\mathbf{X}}{dt} = \frac{d}{dt}(e^{At}\mathbf{X}_0) = A e^{At}\mathbf{X}_0 = A\mathbf{X}.$$

Now $(D^2 - 2D - 3)x_1 = 0$, therefore $x_1 = ae^{3t} + be^{-t}$ and $x_2 = dx_1/dt = 3ae^{3t} - be^{-t}$, where a and b are arbitrary constants. Hence $\mathbf{X}_0 = \begin{pmatrix} a+b \\ 3a-b \end{pmatrix}$. Since $\mathbf{X} = e^{At}\mathbf{X}_0$, we have by inserting the solutions for x_1 and x_2

$$\begin{pmatrix} ae^{3t} + be^{-t} \\ 3ae^{3t} - be^{-t} \end{pmatrix} = e^{At}\begin{pmatrix} a+b \\ 3a-b \end{pmatrix},$$

which gives

$$e^{At} = \tfrac{1}{4}\begin{pmatrix} e^{3t} + 3e^{-t} & e^{3t} - e^{-t} \\ 3e^{3t} - 3e^{-t} & 3e^{3t} + e^{-t} \end{pmatrix}.$$

16.14 Using the transformation $x = t^3$, or otherwise, show that the equation

$$9x^2 \frac{d^2y}{dx^2} + 6x\frac{dy}{dx} + \lambda x^{2/3}y = 0,$$

where λ is a constant, has a non-zero solution satisfying the conditions $y = 0$ when $x = 0$ and when $x = 1$ only if $\lambda = n^2\pi^2$ ($n = 1, 2, 3, \ldots$).

$$\frac{dy}{dx} = \frac{dy}{dt}\frac{dt}{dx}, \quad t = x^{1/3},$$

$$\frac{dt}{dx} = \tfrac{1}{3}x^{-2/3} \quad \text{so} \quad \frac{dy}{dx} = \tfrac{1}{3}x^{-2/3}\frac{dy}{dt}.$$

$$\frac{d^2y}{dx^2} = \frac{1}{3}\frac{d}{dx}\left(x^{-2/3}\frac{dy}{dt}\right) = \frac{1}{3}\left(-\tfrac{2}{3}x^{-5/3}\frac{dy}{dt} + x^{-2/3}\frac{d^2y}{dt^2}\frac{dt}{dx}\right)$$

$$= \frac{1}{3}\left(-\tfrac{2}{3}x^{-5/3}\frac{dy}{dt} + \tfrac{1}{3}x^{-4/3}\frac{d^2y}{dt^2}\right).$$

Hence

$$9x^2\frac{d^2y}{dx^2} = \frac{9}{3}\left(-\tfrac{2}{3}x^{-1/3}\frac{dy}{dt} + \frac{x^{2/3}}{3}\frac{d^2y}{dt^2}\right) = t^2\frac{d^2y}{dt^2} - 2t\frac{dy}{dt},$$

and

$$6x\frac{dy}{dx} = 6x \cdot \tfrac{1}{3}x^{-2/3}\frac{dy}{dt} = 2t\frac{dy}{dt}.$$

Therefore

$$9x^2\frac{d^2y}{dx^2}+6x\frac{dy}{dx}+\lambda x^{2/3}y=t^2\frac{d^2y}{dt^2}-2t\frac{dy}{dt}+2t\frac{dy}{dt}+\lambda t^2y=0,$$

and the equation becomes $d^2y/dt^2+\lambda y=0$, whose solution is

$$y=A\cos\sqrt{\lambda}t+B\sin\sqrt{\lambda}t=A\cos\sqrt{\lambda}x^{1/3}+B\sin\sqrt{\lambda}\,x^{1/3},$$

where A and B are arbitrary constants. The boundary conditions lead to $A=0$, $\sin\sqrt{\lambda}=0$, whence $\sqrt{\lambda}=n\pi$, $(n=1,2,3,\ldots,)$ or $\lambda=n^2\pi^2$.

16.15 (a) By Picard's iterative method, find the first three terms of the solution of the equation

$$\frac{dy}{dx}=\frac{x^2+y^2}{x}$$

for which $y(0)=0$.
(b) Use Picard's method of successive approximation to obtain a power series solution of the system

$$\frac{dy}{dx}=x+z,\frac{dz}{dx}=x-y^2$$

obtaining series for $y(x)$ and $z(x)$ correct to the term in x^2, and taking $y(0)=2$ and $z(0)=1$. Calculate, correct to two decimal places, the values of y and z at $x=0.1$. (L.U.)

Basic result

For the equation $dy/dx=f(x,y)$ with $y(0)=c$, where c is a given constant, the Picard iterative method gives the nth approximation in terms of the $(n-1)$th approximation by the relation

$$y_n=y(0)+\int_0^x f(x,y_{n-1})\,dx.$$

(a) Here $y(0)=0$. Hence

$$y_n=\int_0^x\frac{x^2+y_{n-1}^2}{x}\,dx.$$

Then $y_1=\int_0^x x\,dx=\tfrac12x^2$, taking $y_0=y(0)=0$. Likewise

$$y_2=\int_0^x\frac{x^2+(\tfrac12x^2)^2}{x}\,dx=\tfrac12x^2+\tfrac1{16}x^4,$$

Worked examples in mathematics for scientists and engineers

and

$$y_3 = \int_0^x \frac{x^2 + (\frac{1}{2}x^2 + \frac{1}{16}x^4)^2}{x} \, dx = \frac{1}{2}x^2 + \frac{1}{16}x^4 + \frac{1}{96}x^6 + \frac{1}{2048}x^8.$$

The first three terms of this expression are unchanged in the higher approximations.

(b) Extending the basic result above to a pair of equations, we write

$$\frac{dy}{dx} = f(x, y, z), \quad \frac{dz}{dx} = g(x, y, z), \quad y(0) = c_1, \quad z(0) = c_2,$$

where c_1 and c_2 are given constants.

Then the Picard method gives

$$y_n = y(0) + \int_0^x f(x, y_{n-1}, z_{n-1}) \, dx, \quad z_n = z(0) + \int_0^x g(x, y_{n-1}, z_{n-1}) \, dx.$$

In the case of the given equations, $f(x, y, z) = x + z$, $g(x, y, z) = x - y^2$. Hence

$$y_1 = 2 + \int_0^x (x + 1) \, dx = 2 + \frac{x^2}{2} + x, \quad \text{with} \quad z_0 = z(0) = 1,$$

$$z_1 = 1 + \int_0^x (x - 2^2) \, dx = 1 + \frac{x^2}{2} - 4x \quad \text{with} \quad y_0 = y(0) = 2.$$

Likewise

$$y_2 = 2 + \int_0^x \left(x + 1 - 4x + \frac{x^2}{2} \right) dx = 2 - \frac{3x^2}{2} + x + \frac{x^3}{6},$$

$$z_2 = 1 + \int_0^x \left[x - \left(2 + x + \frac{x^2}{2} \right)^2 \right] dx = 1 - 4x - \frac{3x^2}{2} - x^3 - \frac{x^4}{4} - \frac{x^5}{20}.$$

As far as x^2, the terms are unchanged by going to the next approximation. Hence the required solutions are

$$y = 2 + x - \tfrac{3}{2}x^2, \ z = 1 - 4x - \tfrac{3}{2}x^2,$$

giving $y(0.1) = 2.08$ and $z(0.1) = 0.59$.

16.16 Examine the Picard iterative method in relation to the equation

$$\frac{dy}{dx} = \sqrt{y}, \quad \text{given } y(0) = 0.$$

It is easily seen that (apart from the trivial solution $y = 0$) the equation has the solution $y = x^2/4$ satisfying the given initial condition.

Now Picard's method gives

$$y_n = y(0) + \int_0^x \sqrt{y_{n-1}}\, dx.$$

Hence

$$y_1 = 0 + \int_0^x 0\, dx = 0,$$

$$y_2 = 0 + \int_0^x 0\, dx = 0,$$

and so on. In general, therefore, $y_n = 0$, and the Picard method generates only the trivial solution. A sufficient condition for the Picard method to apply is that $|\partial f/\partial y|$ must be finite in the range of integration. This is known as the Lipschitz condition. Here

$$\frac{\partial f}{\partial y} = \frac{1}{\sqrt{y}}$$

and therefore $|\partial f/\partial y| \to \infty$ as $y \to 0$. The Picard method does not produce the exact non-trivial solution $y = x^2/4$.

16.17 Use the Leibnitz–Maclaurin method to obtain a series solution up to the term in x^4 of the equation

$$x^2 \frac{d^2 y}{dx^2} + x \frac{dy}{dx} - (1-x)y = 0$$

given that $dy/dx = 1$ at $x = 0$.

Differentiating the equation n times by the Leibnitz formula (see 4.8.) we have

$$D^n \left\{ x^2 \frac{d^2 y}{dx^2} + x \frac{dy}{dx} - (1-x)y \right\} = 0,$$

which gives

$$(x^2 D^{n+2} y + 2nx D^{n+1} y + n(n-1)D^n y) + (x D^{n+1} y + n D^n y) - D^n y$$
$$+ (x D^n y + n D^{n-1} y) = 0.$$

At $x = 0$,

$$n(n-1)D^n y + (n-1)D^n y + n D^{n-1} y = 0.$$

Therefore $(n^2 - 1)D^n y = -n D^{n-1} y$ at $x = 0$, and when

$n = 2 : D^2 y = -\frac{2}{3}D(y) = -\frac{2}{3}$, since $dy/dx = 1$ when $x = 0$.

$n = 3 : D^3 y = -\frac{3}{8}D^2 y = -\frac{3}{8}(-\frac{2}{3}) = \frac{1}{4}$.

$n = 4 : D^4 y = -\frac{4}{15}D^3 y = -\frac{4}{15} \cdot \frac{1}{4} = -\frac{1}{15}$.

Hence

$$y(x) = y(0) + xy'(0) + \frac{x^2}{2!} y''(0) + \frac{x^3}{3!} y'''(0) + \cdots$$

$$= y(0) + x - \frac{2}{3} \frac{x^2}{2!} + \frac{1}{4} \frac{x^3}{3!} - \frac{1}{15} \frac{x^4}{4!} + \cdots$$

But from the equation $y = 0$ when $x = 0$. Hence the solution is

$$y = x - \frac{2}{3} \frac{x^2}{2!} + \frac{1}{4} \frac{x^3}{3!} - \frac{1}{15} \frac{x^4}{4!} + \cdots$$

16.18 Find the Frobenius series solutions of the differential equation

$$3x \frac{d^2 y}{dx^2} - (1 - x) \frac{dy}{dx} - y = 0$$

in the form $y = \sum_{n=0}^{\infty} a_n x^{n+c}$, where c is a constant to be found. Show that, if $y(0) = 0$,

$$y = Ax^{4/3} \left(1 - \frac{x}{21} + \frac{x^2}{315} - \cdots \right),$$

where A is an arbitrary constant.

$y = \sum_{n=0}^{\infty} a_n x^{n+c}$. Hence $y' = \sum_{n=0}^{\infty} a_n (n+c) x^{n+c-1}$, $y'' = \sum_{n=0}^{\infty} a_n (n+c)(n+c-1) x^{n+c-2}$. Inserting in the equation we find

$$3 \sum_{n=0}^{\infty} a_n (n+c)(n+c-1) x^{n+c-1} - \sum_{n=0}^{\infty} a_n (n+c) x^{n+c-1}$$
$$+ \sum_{n=0}^{\infty} a_n (n+c) x^{n+c} - \sum_{n=0}^{\infty} a_n x^{n+c} = 0$$

or

$$\sum_{n=0}^{\infty} [3(n+c)(n+c-1) - (n+c)] a_n x^{n+c-1}$$
$$+ \sum_{n=0}^{\infty} a_n (n+c-1) x^{n+c} = 0.$$

Since this must be true for all x, the coefficient of each power of x must be equated to zero.

The lowest power of x occurs when $n = 0$ (this being the x^{c-1} term). Equating this coefficient to zero gives the *indicial* equation for c:

$$3c(c-1) - c = 0 \quad \text{(assuming } a_0 \neq 0\text{)}.$$

Hence

$$c = 0, \tfrac{4}{3}.$$

The coefficient of x^{n+c} is

$$[3(n+c+1)(n+c)-(n+c+1)]a_{n+1}+(n+c-1)a_n,$$

which gives the recurrence relation

$$a_{n+1} = -\frac{(n+c-1)}{(n+c+1)(3n+3c-1)}\,a_n$$

for a_n.
When $c = 0$,

$$a_{n+1} = \frac{-(n-1)}{(n+1)(3n-1)}\,a_n,$$

and therefore

$$a_1 = -a_0, \ a_2 = a_3 = \cdots = a_n = 0.$$

Hence we have the finite series solution $y = a_0(1-x)$.
When $c = 4/3$,

$$a_{n+1} = \frac{-(n+\tfrac{1}{3})}{(n+\tfrac{7}{3})(3n+3)}\,a_n,$$

and we generate the *infinite* series solution

$$y = Ax^{4/3}\left(1-\frac{x}{21}+\frac{x^2}{315}-\cdots\right),$$

where A is an arbitrary constant.

16.19 Find the eigenfunctions and eigenvalues of the equation

$$\frac{d^2\phi}{dx^2}+\lambda\phi = 0,$$

where λ is a parameter, subject to the boundary conditions $\phi'(0)=$ $\phi'(l)=0$, and normalize the eigenfunctions to unity in the range $0 \le x \le l$, where l is a constant.

The solution is $\phi(x)=A\cos\sqrt{\lambda}x + B\sin\sqrt{\lambda}x$, (where A and B are arbitrary constants). Hence $\phi'(x) = -\sqrt{\lambda}A\sin\sqrt{\lambda}x + B\sqrt{\lambda}\cos\sqrt{\lambda}x$. At $x=0$, $\phi'(0)=0$, so $B=0$. Similarly at $x=l$, $\phi'(l)=0$, so $\sin\sqrt{\lambda}l=0$. Hence $\lambda=(n\pi/l)^2$, where $n=0,1,2,3,\ldots$ Accordingly the eigenfunctions are

$$\phi_n = A_n\cos\frac{n\pi x}{l}, \qquad n = 0,1,2,3,\ldots$$

Normalizing to unity requires that

$$\int_0^l \phi_n^2 \, dx = 1,$$

which gives

$$\phi_0 = \frac{1}{\sqrt{l}}, \quad \phi_n = \sqrt{\frac{2}{l}} \cos \frac{n\pi x}{l}, \quad n = 1, 2, 3 \ldots .$$

16.20 Find a series solution for large x for

$$(1 - x^2) \frac{d^2 y}{dx^2} - 2x \frac{dy}{dx} + 2y = 0.$$

Put $x = 1/t$. Then

$$\frac{dy}{dx} = -t^2 \frac{dy}{dt}, \quad \frac{d^2 y}{dx^2} = t^4 \frac{d^2 y}{dt^2} + 2t^3 \frac{dy}{dt}.$$

Hence the equation becomes

$$(t^4 - t^2) \frac{d^2 y}{dx^2} + 2t^3 \frac{dy}{dt} + 2y = 0.$$

Writing $y = \sum_{n=0}^{\infty} a_n t^{n+c}$, the indicial equation becomes (following 16.18) $(c^2 - c - 2)a_0 = 0$. If $a_0 \neq 0$, then $c = 2, -1$. Also we find $a_1[-c(c + 1) + 2] = 0$, and therefore $a_1 = 0$ for both c values. The recurrence relation is

$$a_{n+2} = \frac{n + c + 1}{n + c + 3} a_n.$$

Hence for $c = 2$

$$y = a_0 t^2 (1 + \tfrac{3}{5} t^2 + \tfrac{3}{7} t^4 + \cdots)$$

$$= \frac{a_0}{x^2} \left(1 + \frac{3}{5} \frac{1}{x^2} + \frac{3}{7} \frac{1}{x^4} + \cdots \right),$$

which is the required series.

When $c = -1$ the recurrence relation is

$$a_{n+2} = \frac{n}{n + 2} a_n$$

so, since $a_1 = 0$, $a_3 = a_5 = a_7 = \cdots = a_{2n+1} = 0$. Also $a_2 = a_4 = \cdots = a_{2n} = 0$. Therefore $y = a_0 t^{-1} = a_0 x$ is another solution.

16.21 Show that the equation

$$\frac{d^2y}{dx^2} + P\frac{dy}{dx} + Qy = 0,$$

where P and Q are functions of x, may be put into *normal* form

$$\frac{d^2v}{dx^2} = f(x)v$$

by the substitution $y = uv$, where $u = e^{(-\int P\,dx)/2}$ and

$$f = \left(\frac{P^2}{4} + \frac{P'}{2} - Q\right)$$

Hence solve the equation

$$\frac{d^2y}{dx^2} + x\frac{dy}{dx} + \frac{x^2}{4}\,y = 0.$$

$y = uv$, $y' = uv' + vu'$, $y'' = uv'' + 2u'v' + vu''$. Then $u''v + 2u'v' + uv'' + P(uv' + u'v) + Quv = 0$. Now choose $2u'v' + Puv' = 0$ so that $u = e^{-(\int P\,dx)/2}$. Then finally

$$v'' = \left(\frac{P^2}{4} + \frac{P'}{2} - Q\right)v,$$

which is the required form.

If $P = x$, $Q = x^2/4$ then $f = (x^2/4 + \tfrac{1}{2} - x^2/4) = \tfrac{1}{2}$. The v equation now becomes $v'' = \tfrac{1}{2}v$, which has as its solution $v = A e^{x/\sqrt{2}} + B e^{-x/\sqrt{2}}$, ($A$, B arbitrary). Hence, since $u = e^{-(\int P\,dx)/2} = e^{-(\int x\,dx)/2} = e^{-x^2/4}$, we have

$$y = e^{-x^2/4}(A e^{x/\sqrt{2}} + B e^{-x/\sqrt{2}}).$$

16.22 Show that the (non-linear) Riccati equation

$$\frac{dy}{dx} + y^2 + Py + Q = 0,$$

where P and Q are functions of x, may be transformed to the general linear second order equation

$$\frac{d^2z}{dx^2} + P\frac{dz}{dx} + Qz = 0$$

by the transformation $y = (1/z)\,dz/dx$. Hence solve the Riccati equation with $P = x$ and $Q = x^2/4$.

$$\frac{dy}{dx} = -\frac{1}{z^2}\left(\frac{dz}{dx}\right)^2 + \frac{1}{z}\frac{d^2z}{dx^2}.$$

Inserting in the Riccati equation we find the equation reduces to the given linear equation in z. This can frequently be solved using the Frobenius technique of 16.18.

With $P = x$ and $Q = \dfrac{x^2}{4}$ we have

$$\frac{d^2z}{dx^2} + x\frac{dz}{dx} + \frac{x^2}{4}\,z = 0,$$

which is the equation of 16.21. The solution is

$$z = e^{-x^2/4}(A e^{x/\sqrt{2}} + B e^{-x\sqrt{2}}).$$

Hence

$$y = \frac{1}{z}\frac{dz}{dx} = -\frac{x}{2} + \frac{1}{\sqrt{2}}\left(\frac{A e^{x/\sqrt{2}} - B e^{-x/\sqrt{2}}}{A e^{x/\sqrt{2}} + B e^{-x/\sqrt{2}}}\right).$$

16.23 Show that Pinney's equation

$$\frac{d^2y}{dx^2} + p(x)y = \frac{c}{y^3}$$

is satisfied by

$$y = \left[u^2 + \frac{c}{W^2}\,v^2\right]^{1/2},$$

where u and v are independent solutions of $d^2z/dx^2 + p(x)z = 0$, and $W = uv' - vu'$ ($=$ constant), c being an arbitrary constant.

We first write $c/W^2 = \lambda$ (say). Now since u and v satisfy $z'' + p(x)z = 0$ then $u'' = -pu$ and $v'' = -pv$. Clearly $u''/u = v''/v$, so $u''v - v''u = 0$ or $d/dx\ (uv' - u'v) = 0$, whence $W = $ constant. Now putting $y = (u^2 + \lambda v^2)^{1/2}$ into the Pinney equation and using $u'' = -pu$, $v'' = -pv$ we find, after some manipulation, that $\lambda(uv' - vu')^2 = c$. But, since $uv' - vu' = W$, this is identically satisfied. Hence the given form of y satisfies Pinney's equation.

It is worth giving an example here. Suppose we take $p(x) = 1$. Then two independent solutions of $z'' + z = 0$ are $\sin x$ and $\cos x$. Hence $u = \sin x$, $v = \cos x$, $uv' - vu' = W = -1$. Therefore

$$y = (\sin^2 x + c \cos^2 x)^{1/2}$$

is a solution of the Pinney equation

$$y'' + y = \frac{c}{y^3}.$$

17. SPECIAL FUNCTIONS

[Bessel functions, Liouville–Green method, Hermite polynomials, Legendre polynomials, Airy equation]

17.1 Obtain the solution (Bessel's function of zero order)

$$y = J_0(x) = 1 - \frac{x^2}{2^2} + \frac{x^4}{2^2 \cdot 4^2} - \cdots + \frac{(-1)^n x^{2n}}{2^{2n}(n!)^2} + \cdots$$

of the differential equation

$$x\frac{d^2y}{dx^2} + \frac{dy}{dx} + xy = 0.$$

Show that $y = uJ_0(x)$ is a second solution if $u = \int \frac{dx}{xJ_0^2(x)}$, and expand u in the form $a + \log_e x + b_2 x^2 + \cdots$, where a is arbitrary and b_2 is to be found.

Using the Leibnitz–Maclaurin method (see 16.17.)

$$D^n\left(x\frac{d^2y}{dx^2}\right) = xD^{n+2}y + nD^{n+1}, \qquad D^n\left(\frac{dy}{dx}\right) = D^{n+1}y.$$

Therefore

$$xD^{n+2}y + nD^{n+1}y + D^{n+1}y + xD^ny + nD^{n-1}y = 0.$$

Hence at $x = 0$, $D^{n+1}y = -(n/n+1)D^{n-1}y$, and therefore when

$n = 1$: $D^2y = -\frac{1}{2}y(0)$

$n = 2$: $D^3y = -\frac{2}{3}y'(0) = 0$ (since $y'(0) = 0$ from the equation itself)

$n = 3$: $D^4y = -\frac{3}{4}y''(0) = \frac{1 \cdot 3}{4 \cdot 2}y(0)$

$n = 4$: $D^5y = 0$ (since $y'''(0) = 0$)

$n = 5$: $D^6y = -\frac{5}{6}y''''(0) = \frac{1 \cdot 3 \cdot 5}{6 \cdot 4 \cdot 2}y(0).$

Therefore

$$y = y(0)\left(1 - \frac{x^2}{2^2} + \frac{x^4}{2^2 \cdot 4^2} - \cdots\right).$$

The Bessel function of zero order $J_0(x)$ is this series with $y(0)$ chosen to be unity. Now let $y = uJ_0(x)$ so that the equation becomes

$$x(uJ_0'' + 2u'J_0' + u''J_0) + (uJ_0' + u'J_0) + xuJ_0 = 0.$$

Hence

$$u[xJ_0'' + J_0' + xJ_0] + xu''J_0 + u'(2xJ_0' + J_0) = 0.$$

Since Bessel's equation is satisfied by J_0 the first term is zero. Hence

$$xu'' + u'\left(2x\frac{J_0'}{J_0} + 1\right) = 0$$

and therefore

$$\frac{u''}{u'} + 2\frac{J_0'}{J_0} + \frac{1}{x} = 0.$$

Integrating gives

$$\log_e u' + \log_e J_0^2 + \log_e x = \log_e A$$
$$\text{(where } A \text{ is an arbitrary constant),}$$

whence

$$u'xJ_0^2 = A.$$

Therefore

$$u = A\int \frac{\mathrm{d}x}{xJ_0^2},$$

where A may be chosen to be unity without loss of generality. Now

$$J_0^2 = \left(1 - \frac{x^2}{2^2} + \frac{x^4}{2^24^2} - \cdots\right)^2 = \left(1 - \frac{x^2}{2} + \text{higher powers}\right).$$

Therefore

$$J_0^{-2} = \left(1 + \frac{x^2}{2} + \text{higher powers}\right).$$

Hence

$$u = \int \frac{1}{x}\left(1 + \frac{x^2}{2} + \cdots\right)\mathrm{d}x = \log_e x + \frac{x^2}{4} + \cdots + a,$$

where a is an arbitrary constant. Hence $b_2 = \frac{1}{4}$.

17.2 Show that $\dfrac{1}{x^n} J_n(x)$ is a solution of

$$\frac{d^2y}{dx^2} + \frac{1+2n}{x}\frac{dy}{dx} + y = 0,$$

and that $\sqrt{x}\, J_n(kx)$ (k constant) is a solution of

$$\frac{d^2y}{dx^2} + \left(k^2 - \frac{4n^2-1}{4x^2}\right)y = 0,$$

where in both cases n is a positive integer, and J_n is Bessel's function of order n which is a solution of the Bessel equation

$$x^2 \frac{d^2y}{dx^2} + x\frac{dy}{dx} + (x^2 - n^2)y = 0.$$

$$\frac{d}{dx}\left(\frac{J_n}{x^n}\right) = \frac{J_n'}{x^n} - \frac{n}{x^{n+1}} J_n$$

$$\frac{d^2}{dx^2}\left(\frac{J_n}{x^n}\right) = \frac{J_n''}{x^n} - \frac{2n}{x^{n+1}} J_n' + \frac{n(n+1)}{x^{n+2}} J_n.$$

Therefore equation becomes

$$\frac{J_n''}{x^n} - \frac{2n}{x^{n+1}} J_n' + \frac{n(n+1)}{x^{n+2}} J_n + \left(\frac{1+2n}{x}\right)\left(\frac{J_n'}{x^n} - \frac{n}{x^{n+1}} J_n\right) + \frac{J_n}{x^n}$$

$$= \frac{1}{x^n}\left(J_n'' + \frac{1}{x} J_n' + J_n\right) - \frac{n^2}{x^{n+2}} J_n$$

$$= (x^2 J_n'' + xJ_n + (x^2 - n^2)J_n)/x^{n+2} = 0,$$

since J_n is a solution of Bessel's equation. Hence $(1/x^n)J_n(x)$ is a solution of the first (given) equation. Now

$$y = \sqrt{x}J_n(kx),\quad y' = \sqrt{x}kJ_n'(kx) + \frac{1}{2\sqrt{x}} J_n(kx),$$

$$y'' = \sqrt{x}k^2 J_n''(kx) + \frac{k}{\sqrt{x}} J_n'(kx) - \frac{1}{4x^{3/2}} J_n(kx).$$

Therefore the second (given) equation becomes

$$y'' + \left(k^2 - \frac{4n^2-1}{x^2}\right)y = \sqrt{x}k^2 J_n'' + \frac{k}{\sqrt{x}} J_n'$$

$$- \frac{1}{4x^{3/2}} J_n + k^2\sqrt{x}J_n - \frac{(4n^2-1)}{4x^{3/2}} J_n$$

$$= \frac{1}{x^{3/2}}[k^2 x^2 J_n'' + kxJ_n' + (k^2 x^2 - n^2)J_n],$$

(primes denoting differentiation with respect to the argument of the function) which is zero since $J_n(kx)$ is a solution of

$$k^2 x^2 J_n'' + kx J_n' + (k^2 x^2 - n^2) J_n = 0.$$

17.3 Use the result of 17.2 to show that the general solution of Bessel's equation for large positive values of x is

$$J_n(x) \sim (A \cos x + B \sin x)/x^{1/2},$$

where A and B are arbitrary constants.

From 17.2 $\sqrt{x} J_n(x)$ is a solution of

$$\frac{d^2 y}{dx^2} + \left(1 - \frac{4n^2 - 1}{4x^2}\right) y = 0.$$

Hence for large x the solutions are approximately those of $d^2 y/dx^2 + y = 0$ (neglecting the $(4n^2 - 1)/4x^2$ term). Hence $y = A \cos x + B \sin x$, and therefore

$$J_n(x) \sim \frac{1}{\sqrt{x}} (A \cos x + B \sin x).$$

17.4 By making the Liouville–Green transformation

$$x = x(\xi), \qquad G = (\xi')^{1/2} y,$$

where $\xi' = d\xi/dx$, show that the equation

$$\frac{d^2 y}{dx^2} = f(x) y$$

becomes

$$\frac{d^2 G}{d\xi^2} = \left\{\frac{f(x)}{\xi'^2} + \Delta(x(\xi))\right\} G,$$

where $\Delta(x(\xi))$ is the Schwarzian derivative defined by

$$\Delta(x(\xi)) = \frac{\xi'''}{2\xi'^3} - \frac{3}{4} \frac{\xi''^2}{\xi'^4}.$$

Hence solve the equation $d^2 y/dx^2 = e^{nx} y$, $(n \neq 0)$, in the range $-\infty < x < \infty$.

Now

$$y = (\xi')^{-1/2} G, \qquad y' = (\xi')^{1/2} G' - \tfrac{1}{2} \xi'' (\xi')^{-3/2} G,$$

$$y'' = (\xi')^{3/2} G'' + \tfrac{1}{2} (\xi')^{-1/2} \xi'' G' - \tfrac{1}{2} (\xi')^{-1/2} \xi'' G' + \tfrac{3}{4} (\xi')^{-5/2} (\xi'')^2 G \\ - \tfrac{1}{2} \xi''' (\xi')^{-3/2} G.$$

Hence the equation $y'' = f(x)y$ becomes

$$G''(\xi')^{3/2} + \tfrac{3}{4}(\xi')^{-5/2}(\xi'')^2 G - \tfrac{1}{2}\xi'''(\xi')^{-3/2}G - f(x)(\xi')^{-1/2}G = 0$$

or

$$G'' = \left\{\frac{f(x)}{\xi'^2} + \frac{1}{2}\frac{\xi'''}{\xi'^3} - \frac{3}{4}\frac{\xi''}{\xi'^4}\right\}G.$$

Now $y'' = e^{nx}y$ transforms into

$$G'' = \left\{\frac{e^{nx}}{\xi'^2} + \Delta(x(\xi))\right\}G.$$

Choosing $\xi'^2 = e^{nx}$ so that $\xi = (2/|n|)e^{nx/2}$, where $0 < \xi < \infty$, and calculating the higher derivatives, we find

$$\Delta(x(\xi)) = -\frac{n^2}{16e^{nx}} = -\frac{1}{4\xi^2}.$$

Hence finally the equation transforms into

$$\frac{d^2G}{d\xi^2} = \left(1 - \frac{1}{4\xi^2}\right)G.$$

This is a standard equation whose solutions are

$$G = \begin{cases} \xi^{1/2}I_0(\xi), \\ \xi^{1/2}K_0(\xi), \end{cases}$$

where I_0 and K_0 are the *modified* Bessel functions of zero order satisfying the equation $xy'' + y' - xy = 0$. The graphs of these functions are shown in Fig. 29. Substituting back for G and ξ gives y.

17.5 Show that $y = x^{1/2}J_{1/3}(ax^b)$ satisfies Airy's equation $d^2y/dx^2 + xy = 0$ if $a = \tfrac{2}{3}$, $b = \tfrac{3}{2}$.

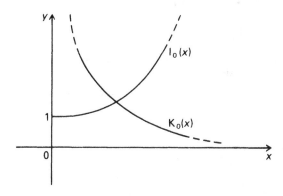

Fig. 29

Put $u = ax^b$. Then

$$y' = abx^{b-1/2}J'_{1/3} + \tfrac{1}{2}x^{-1/2}J_{1/3},$$

$$y'' = a^2b^2x^{2b-3/2}J''_{1/3} + ab(b-\tfrac{1}{2})x^{b-3/2}J'_{1/3}$$

$$-\tfrac{1}{4}x^{3/2}J_{1/3} + \tfrac{1}{2}x^{-1/2}abx^{b-1}J'_{1/3},$$

where primes denote differentiation with respect to u. Hence the equation becomes

$$a^2b^2x^{2b}J''_{1/3} + ab^2x^bJ'_{1/3} + J_{1/3}(x^2 - \tfrac{1}{4}) = 0$$

or, in terms of u,

$$u^2J''_{1/3} + uJ'_{1/3} + \left[\frac{1}{b^2}\left(\frac{u}{a}\right)^{3/b} - \frac{1}{4b^2}\right]J_{1/3} = 0.$$

If $a = \tfrac{2}{3}$, $b = \tfrac{3}{2}$ this becomes

$$u^2J''_{1/3} + uJ'_{1/3} + (u^2 - \tfrac{1}{9})J_{1/3} = 0,$$

which is satisfied identically since it is Bessel's equation with $n = \tfrac{1}{3}$ (see 17.2).

17.6 The Hermite polynomials $H_n(x)$ are defined by

$$H_n(x) = (-1)^n e^{x^2}\frac{d^n}{dx^n}(e^{-x^2}) \qquad (n = 0, 1, 2, 3, \ldots).$$

Show that $H_0 = 1$, $H_1 = 2x$, $H_2 = 4x^2 - 2$, $H_3 = 8x^3 - 12x$, and verify that each satisfies the Hermite differential equation

$$\frac{d^2y}{dx^2} - 2x\frac{dy}{dx} + 2ny = 0$$

with the appropriate value of n. Verify the general result

$$I_{m,n} = \int_{-\infty}^{\infty} e^{-x^2}H_m(x)H_n(x)\,dx = 0 \qquad (m \neq n),$$

when $m = 1$ and $n = 2$. Show that

$$e^{2tx - t^2} = \sum_{n=0}^{\infty} H_n(x)\frac{t^n}{n!}$$

for the specific values $n = 0, 1, 2, 3$. (This is a *generating* function for the Hermite polynomials).

$$H_0 = 1, \qquad H_1 = (-1)^1 e^{x^2}\frac{d}{dx}(e^{-x^2}) = 2x,$$

$$H_2 = (-1)^2 e^{x^2}\frac{d^2}{dx^2}(e^{-x^2}) = 4x^2 - 2,$$

$$H_3 = (-1)^3 e^{x^2}\frac{d^3}{dx^3}(e^{-x^2}) = 8x^3 - 12x.$$

The differential equation is satisfied by each of these functions in turn. For example, when $n = 3$, the equation becomes

$$\frac{d^2}{dx^2}(8x^3 - 12x) - 2x\frac{d}{dx}(8x^3 - 12x) + 6(8x^3 - 12x)$$

$$= 48x - 2x(24x^2 - 12) + 6(8x^3 - 12x) = 0 \text{ identically.}$$

Also

$$I_{1,2} = \int_{-\infty}^{\infty} e^{-x^2}2x(4x^2 - 2)\,dx = 8\int_{-\infty}^{\infty} x^3 e^{-x^2}\,dx - 4\int_{-\infty}^{\infty} xe^{-x^2}\,dx = 0,$$

since each integrand is an odd function. Similarly for other m, n values.

Now

$$H_0 + H_1 t + H_2\frac{t^2}{2!} + H_3\frac{t^3}{3!} + \cdots = e^{2tx - t^2}$$

$$= 1 + (2tx - t^2) + \frac{(2tx - t^2)^2}{2!} + \frac{(2tx - t^2)^3}{3!} + \cdots$$

$$= 1 + (2x)t + (2x^2 - 1)t^2 + \frac{(4x^3 - 6x)}{3}t^3 + \cdots$$

Comparing coefficients of powers of t on each side of this equation we have the given forms of H_0, H_1, H_2, H_3.

17.7 The Legendre polynomials $P_n(x)$ are defined by the Rodrigues formula to be

$$P_n(x) = \frac{1}{2^n n!}\frac{d^n}{dx^n}[(x^2 - 1)^n].$$

Obtain P_0, P_1, P_2 and P_3. Show that $P_n(x)$ satisfies the Legendre equation

$$(1 - x^2)\frac{d^2 y}{dx^2} - 2x\frac{dy}{dx} + n(n + 1)y = 0.$$

$$P_0 = 1, \quad P_1 = \tfrac{1}{2}\cdot 2x = x, \quad P_2 = \frac{1}{2^2 2!}\frac{d^2}{dx^2}(x^2 - 1)^2 = \tfrac{1}{2}(3x^2 - 1),$$

$$P_3 = \frac{1}{2^3 3!}\frac{d^3}{dx^3}(x^2 - 1)^3 = \tfrac{1}{2}(5x^3 - 3x).$$

Let $z = (1/2^n n!)(x^2 - 1)^n$. Then $(x^2 - 1)Dz = 2nxz$. Differentiating $n + 1$ times (using Leibnitz)

$$(x^2 - 1)D^{n+2}z + 2x(n + 1)D^{n+1}z + n(n + 1)D^n z$$
$$= 2n(xD^{n+1}z + (n + 1)D^n z).$$

Hence

$$(1-x^2)D^{n+2}z - 2xD^{n+1}z + [2n(n+1) - n(n+1)]D^n z = 0$$

which is $\qquad (1-x^2)D^{n+2}z - 2xD^{n+1}z + n(n+1)D^n z = 0,$

and therefore

$$(1-x^2)D^2 y - 2xDy + n(n+1)y = 0,$$

where $y = D^n z$. Hence $P_n(x)$ satisfies the Legendre equation.

17.8 By the substitution $P(\theta) = Q(\theta)(\sin\theta)^{-1/2}$ in Legendre's equation

$$\frac{1}{\sin\theta}\frac{d}{d\theta}\left(\sin\theta\frac{dP}{d\theta}\right) + n(n+1)P = 0,$$

show that, when $\csc\theta \ll (2n+1)$, the solutions of Legendre's equation are of the form

$$(\sin\theta)^{-1/2}\begin{cases}\sin(n+\tfrac12)\theta \\ \cos(n+\tfrac12)\theta.\end{cases}$$

$$\frac{dP}{d\theta} = (\sin\theta)^{-1/2}\frac{dQ}{d\theta} - \tfrac12(\sin\theta)^{-3/2}\cos\theta\,Q,$$

$$\sin\theta\frac{dP}{d\theta} = (\sin\theta)^{1/2}\frac{dQ}{d\theta} - \frac12\frac{\cos\theta}{(\sin\theta)^{1/2}}Q,$$

$$\frac{d}{d\theta}\left(\sin\theta\frac{dP}{d\theta}\right) = (\sin\theta)^{1/2}\frac{d^2Q}{d\theta^2}$$

$$+\tfrac12(\sin\theta)^{-1/2}\cos\theta\frac{dQ}{d\theta} - \frac12\frac{\cos\theta}{(\sin\theta)^{1/2}}\frac{dQ}{d\theta}$$

$$-\tfrac12 Q\left[-\frac{\sin\theta}{(\sin\theta)^{1/2}} - \frac12\frac{\cos^2\theta}{(\sin\theta)^{3/2}}\right],$$

$$\frac{1}{\sin\theta}\frac{d}{d\theta}\left(\sin\theta\frac{dP}{d\theta}\right) = \frac{1}{(\sin\theta)^{1/2}}\frac{d^2Q}{d\theta^2}$$

$$+\frac12\frac{1}{(\sin\theta)^{1/2}}\left[1+\frac{1-\sin^2\theta}{2\sin^2\theta}\right]Q.$$

The equation becomes

$$\frac{d^2Q}{d\theta^2}+\left[n(n+1)+\frac12-\frac{1-\sin^2\theta}{4\sin^2\theta}\right]Q = 0$$

which can be written as

$$\frac{d^2Q}{d\theta^2}+[(n+\tfrac{1}{2})^2+\tfrac{1}{4}\cosec^2\theta]Q=0.$$

Hence if

$$\cosec^2\theta \ll 4(n+\tfrac{1}{2})^2=(2n+1)^2$$

that is, if

$$\cosec\theta \ll (2n+1),$$

then

$$Q=A\sin(n+\tfrac{1}{2})\theta+B\sin(n+\tfrac{1}{2})\theta,$$

where A and B are arbitrary constants. Hence finally

$$P=(\sin\theta)^{-1/2}\begin{cases}\sin(n+\tfrac{1}{2})\theta\\\cos(n+\tfrac{1}{2})\theta.\end{cases}$$

17.9   Show that the non-linear equation $y''=xy+2y^3+\tfrac{1}{2}$ has $y'=x/2+y^2$ as a first integral. Hence transform this equation into the Airy equation and obtain solutions of the non-linear equation in terms of Airy functions.

From $y'=x/2+y^2$, we have

$$y''=\tfrac{1}{2}+2yy'=\tfrac{1}{2}+2y\left(\frac{x}{2}+y^2\right)=xy+2y^3+\tfrac{1}{2}.$$

Hence $y'=x/2+y^2$ is a first integral. To solve this equation (which is of Ricatti type—see 16.22) we put

$$y=-\frac{1}{z}\frac{dz}{dx}.$$

Then the equation becomes

$$\frac{d^2z}{dx^2}+\frac{xz}{2}=0,$$

which, on writing $x=2^{1/3}x'$, has the Airy equation form $d^2z/dx'^2+x'z=0$, whose solutions are Ai $(-x')$ and Bi $(-x')$ (the Airy functions). Hence the original equation has solutions.

$$y=-\frac{1}{\text{Ai}(-2^{-1/3}x)}\frac{d}{dx}(\text{Ai}(-2^{-1/3}x)),$$

and

$$y=-\frac{1}{\text{Bi}(-2^{-1/3}x)}\frac{d}{dx}(\text{Bi}(-2^{-1/3}x)).$$

18. LAPLACE TRANSFORM

[Elementary transforms, inverses, Heaviside step function, ordinary differential equations, Bessel functions, Dirac delta function, convolution theorem, Riemann zeta function]

18.1 Find the Laplace transforms of the following functions:
(a) $\sin^2 x$, (b) $x^3 e^{-3x}$, (c) $e^x \sin^2 x$, (d) $x^{1/2}$.

The Laplace transform of a function $f(x)$ is defined as

$$L\{f(x)\} = \bar{f}(p) = \int_0^\infty f(x) e^{-px} \, dx.$$

(a) $L\{\sin^2 x\} = \displaystyle\int_0^\infty e^{-px} \sin^2 x \, dx = \dfrac{1}{2p} - \dfrac{p}{2(p^2 + 4)}$
(integrating by parts).

(b) $L\{x^3 e^{-3x}\} = \displaystyle\int_0^\infty x^3 e^{-(p+3)x} \, dx = \dfrac{6}{(p+3)^4}$ (again by parts).

Basic result

The shift theorem states that

$$L\{e^{-ax}f(x)\} = \int_0^\infty e^{-ax} f(x) e^{-px} \, dx = \bar{f}(p + a).$$

(c) $L\{e^x \sin^2 x\} = \dfrac{1}{2(p-1)} - \dfrac{(p-1)}{2(p-1)^2 + 8}$ (using the result of (a) and the shift theorem).

(d) $L\{x^{1/2}\} = \displaystyle\int_0^\infty x^{1/2} e^{-px} \, dx$. Put $px = u^2$, $p \, dx = 2u \, du$.

Then

$$L\{x^{1/2}\} = \dfrac{2}{p^{3/2}} \int_0^\infty u^2 e^{-u^2} \, du = \dfrac{1}{2} \dfrac{\sqrt{\pi}}{p^{3/2}},$$

using the result $\int_0^\infty u^2 e^{-u^2} du = \sqrt{\pi}/4$ (obtained by differentiating the standard integral $\int_0^\infty e^{-au^2} du = \frac{1}{2}\sqrt{\pi/a}$ with respect to a and then letting $a \to 1$).

18.2 Find inverse Laplace transforms of

(a) $\dfrac{p^2}{(p^2+1)(p^2+2)}$,

(b) $\dfrac{1}{p(p-1)^3}$.

(a) Using partial fractions

$$\frac{p^2}{(p^2+1)(p^2+2)} = \frac{Ap+B}{p^2+1} + \frac{Cp+D}{p^2+2}.$$

Hence comparing powers of p in the numerators on each side we have

$$A+C=0, \qquad B+D=1, \qquad 2B+D=0, \qquad 2A+C=0,$$

which gives

$$A=0, B=-1, C=0, D=2.$$

Hence

$$\frac{p^2}{(p^2+1)(p^2+2)} = -\frac{1}{p^2+1} + \frac{2}{p^2+2}$$

$$L^{-1}\left\{\frac{p^2}{(p^2+1)(p^2+2)}\right\} = -\sin x + \sqrt{2}\sin\sqrt{2}x.$$

(b) In the same way

$$\frac{1}{p(p-1)^3} = \frac{A}{p} + \frac{B}{p-1} + \frac{C}{(p-1)^2} + \frac{D}{(p-1)^3}$$

whence

$$A=-1, B=1, C=-1, D=1.$$

Therefore

$$\frac{1}{p(p-1)^3} = -\frac{1}{p} + \frac{1}{p-1} - \frac{1}{(p-1)^2} + \frac{1}{(p-1)^3}$$

and

$$L^{-1}\left\{\frac{1}{p(p-1)^3}\right\} = -1 + e^x - xe^x + \tfrac{1}{2}x^2 e^x,$$

using the result that

$$x^n e^x = L^{-1} \left\{ \frac{n!}{(p-1)^{n+1}} \right\}.$$

18.3 Show that the Laplace transform of the unit step function (the Heaviside function) defined by

$$H(x-a) = \begin{cases} 0, & x < a \\ 1, & x \geq a \end{cases}, \quad \text{where } a > 0,$$

is

$$L\{H(x-a)\} = e^{-ap}/p.$$

Given that

$$f(x) = k[H(x) - 2H(x-a) - 2H(x-2a) - 2H(x-3a) + \cdots],$$

where k is a constant, show that

$$L\{f(x)\} = \frac{k}{p} \tanh \frac{ap}{2}.$$

$$L\{H(x-a)\} = \int_0^\infty H(x-a)e^{-px} \, dx$$

$$= \int_0^a 0.e^{-px} \, dx + \int_a^\infty 1.e^{-px} \, dx = \frac{e^{-pa}}{p}.$$

$$L\{f(x)\} = k \left[\frac{1}{p} - 2\frac{e^{-pa}}{p} + 2\frac{e^{-2pa}}{p} - 2\frac{e^{-3ap}}{p} + \cdots \right]$$

$$= \frac{k}{p}[1 - 2e^{-pa} + 2e^{-2ap} - 2e^{-3ap} + \cdots]$$

$$= \frac{k}{p}[1 - 2e^{-pa}(1 - e^{-ap} + e^{-2ap} - e^{-3ap} + \cdots)]$$

$$= \frac{k}{p}[1 - 2e^{-ap}(1 + e^{-ap})^{-1}]$$

(using the binomial expansion of $(1 + e^{-ap})^{-1}$, which is valid since $|e^{-ap}| < 1$). Hence

$$L\{f(x)\} = \frac{k}{p} \left[1 - \frac{2e^{-ap}}{1 + e^{-ap}} \right] = \frac{k}{p} \left[\frac{1 - e^{-ap}}{1 + e^{-ap}} \right] = \frac{k}{p} \tanh \frac{ap}{2}.$$

18.4 Solve the following equations by the Laplace transform method:

(a) $(D^2 + 4D + 8)y = \cos 2x$

given $y = 2$ and $dy/dx = 1$ at $x = 0$.

(b) $(D+1)y+Dz=0$,

$(D-1)y+2Dz=e^{-x}$,

given $y=\tfrac{1}{2}$ and $z=0$ at $x=0$.
(In both cases $D\equiv d/dx$).

Basic result

$$L\left\{\frac{dy}{dx}\right\}=p\bar{y}(p)-y(0),\qquad L\left\{\frac{d^2y}{dx^2}\right\}=p^2\bar{y}(p)-py(0)-y'(0),$$

where $y(0)$ and $y'(0)$ are respectively the function and its derivative at $x=0$.
(a) Taking the Laplace transform of the equation and using the above basic results we have

$$p^2\bar{y}(p)-py(0)-y'(0)+4(p\bar{y}(p)-y(0))+8\bar{y}(p)=\frac{p}{p^2+4}.$$

Therefore

$$(p^2+4p+8)\bar{y}(p)=2p+9+\frac{p}{p^2+4}.$$

Hence

$$\bar{y}(p)=\frac{2p+9}{p^2+4p+8}+\frac{p}{(p^2+4)(p^2+4p+8)}.$$

Using partial fractions (as in 18.2) we finally obtain

$$y(x)=L^{-1}\{\bar{y}(p)\}=\tfrac{1}{10}\sin 2x+\tfrac{1}{20}\cos 2x$$
$$+\tfrac{1}{20}e^{-2x}(39\cos 2x+47\sin 2x).$$

(b) Let $\bar{y}(p)$ be the transform of $y(x)$ and $\bar{z}(p)$ be the transform of $z(x)$. Then

$$p\bar{y}(p)-y(0)+\bar{y}(p)+p\bar{z}(p)-z(0)=0,$$

$$p\bar{y}(p)-y(0)-\bar{y}(p)+2p\bar{z}(p)-2z(0)=\frac{1}{p+1}.$$

Therefore

$$p\bar{y}(p)-\tfrac{1}{2}+\bar{y}(p)+p\bar{z}(p)=0,$$

$$p\bar{y}(p)-\tfrac{1}{2}-\bar{y}(p)+2p\bar{z}(p)=\frac{1}{p+1}.$$

143

Solving for $\bar{y}(p)$ gives

$$\bar{y}(p) = \frac{1}{2(p+3)} - \frac{1}{(p+1)(p+3)} = \frac{1}{p+3} - \frac{1}{2(p+1)}$$

and hence

$$y(x) = L^{-1}\{\bar{y}(p)\} = e^{-3x} - \tfrac{1}{2}e^{-x}.$$

Either by solving for $\bar{z}(p)$ and inverting in the same way, or by using the solution for $y(x)$ and inserting it into the first equation which links $y(x)$ and $z(x)$, we find

$$z(x) = \tfrac{2}{3}(1 - e^{-3x}).$$

18.5 Given that the zero-order Bessel function $J_0(x)$ is defined by

$$\sum_{n=0}^{\infty} \frac{(-1)^n}{(n!)^2}\left(\frac{x}{2}\right)^{2n}$$

(see 17.1), prove that

$$L\{J_0(x)\} = \frac{1}{\sqrt{p^2+1}}.$$

Deduce the transform of $J_0(ax)$, where a is a positive constant.

Now $L\{x^n\} = n!/p^{n+1}$, where n is a positive integer. Therefore

$$L\{J_0(x)\} = L\left\{1 - \frac{x^2}{2^2} + \frac{x^4}{2^2 4^2} - \frac{x^6}{2^2 4^2 6^2} + \cdots\right\}$$

$$= \frac{1}{p} - \frac{1}{2^2}\frac{2!}{p^3} + \frac{1}{2^2 4^2}\frac{4!}{p^5} - \cdots = \frac{1}{p}\left(1 + \frac{1}{p^2}\right)^{-1/2} = \frac{1}{\sqrt{p^2+1}}.$$

Also in general

$$L\{f(ax)\} = \int_0^\infty f(ax)e^{-px}\,dx = \frac{1}{a}\int_0^\infty f(u)e^{-pu/a}\,du = \frac{1}{a}\bar{f}\left(\frac{p}{a}\right).$$

Therefore

$$L\{J_0(ax)\} = \frac{1}{a}\frac{1}{\sqrt{(p/a)^2+1}} = \frac{1}{\sqrt{p^2+a^2}}.$$

18.6 Show that if $L\{f(x)\} = \bar{f}(p)$, then $L\{xf(x)\} = -d\bar{f}(p)/dp$. Hence deduce that $L\{x^{3/2}\} = \tfrac{3}{4}\sqrt{\pi/p^5}$ using the result of 18.1 for $L\{x^{1/2}\}$.

$$L\{x^{1/2}\} = \tfrac{1}{2}\sqrt{\pi/p^3}.$$

Hence

$$L\{x^{3/2}\} = -\frac{d}{dp}\left(\frac{1}{2}\sqrt{\frac{\pi}{p^3}}\right) = \frac{3}{4}\sqrt{\frac{\pi}{p^5}}.$$

18.7 Evaluate $L\{xH(x-a)\}$, where $a>0$.

$$L\{xH(x-a)\} = -\frac{d}{dp} L\{H(x-a)\} \quad \text{(from 18.6)}.$$

$$L\{H(x-a)\} = \frac{e^{-ap}}{p} \quad \text{(see 18.3)}.$$

Hence

$$L\{xH(x-a)\} = -\frac{d}{dp}\left(\frac{e^{-ap}}{p}\right) = \frac{(1+ap)}{p^2} e^{-ap}.$$

18.8 Show that the transform of the function $H(x-1)\cos(x-1)$ is

$$\frac{pe^{-p}}{p^2+1}.$$

Hence solve by the Laplace transform method the differential equation

$$\frac{d^2y}{dx^2} + y = H(x-1)$$

given $y(0)=0$, $y'(0)=0$.

$$L\{H(x-1)\cos(x-1)\} = \int_1^\infty \cos(x-1)e^{-px}\,dx \quad (\text{ since } H(x-1)$$
$$= 0 \text{ for } 0 \leqslant x < 1)$$
$$= \frac{pe^{-p}}{p^2+1} \quad \text{(by integration by parts)}.$$

Transforming the equation gives

$$p^2\bar{y}(p) - py(0) - y'(0) + \bar{y}(p) = \frac{e^{-p}}{p}. \text{ Using the initial conditions}$$

$$\bar{y}(p) = \frac{e^{-p}}{p(p^2+1)} = e^{-p}\left(\frac{1}{p} - \frac{p}{p^2+1}\right)$$

and

$$y(x) = L^{-1}\{\bar{y}(p)\} = H(x-1) - H(x-1)\cos(x-1)$$
$$= H(x-1)(1-\cos(x-1)).$$

18.9 Verify the following properties of the Dirac delta function:

(a) $\delta(-x) = \delta(x)$,
(b) $2\delta(2x) = \delta(x)$,
(c) $x\delta(x) = 0$,
(d) $f(x)\delta(x-a) = f(a)\delta(x-a)$ (a constant).

Basic result

The δ-function is defined by

$$\int_{-\infty}^{\infty} \delta(x)\, dx = 1, \qquad \delta(x) = 0, \quad x \neq 0.$$

We note that $\delta(x)$ has the sieving property of picking out the function at a particular value of x since, from the definition,

$$\int_{-\infty}^{\infty} f(x)\delta(x-a)\, dx = f(a),$$

where $f(x)$ is some continuous function, and a is a constant.

(a) $\displaystyle \int_{-\infty}^{\infty} \delta(-x)\, dx = \int_{-\infty}^{\infty} \delta(x')\, dx' = 1$, where $x' = -x$. Also $\delta(x) = 0$, $x \neq 0$, implies that $\delta(-x) = 0$, $x \neq 0$. Hence $\delta(-x)$ satisfies the same requirements as $\delta(x)$, and therefore $\delta(-x) = \delta(x)$.

(b) $\displaystyle \int_{-\infty}^{\infty} 2\delta(2x)\, dx = \int_{-\infty}^{\infty} \delta(2x)\, d(2x) = \int_{-\infty}^{\infty} \delta(u)\, du = 1$,

where $2x$ has been written as u. Hence $2\delta(2x)$ is the same as $\delta(x)$.

(c) $\displaystyle \int_{-\infty}^{\infty} x\delta(x)\, dx = 0$, since for $x \neq 0$ the integrand is zero in virtue of $\delta(x)$ being zero, and for $x = 0$ the integrand is zero anyway. Hence

$$x\,\delta(x) = 0.$$

(d) From the basic result we have $\int_{-\infty}^{\infty} f(x)\delta(x-a)\, dx = f(a)$. On the other hand

$$\int_{-\infty}^{\infty} f(a)\, \delta(x-a)\, dx = f(a)\int_{-\infty}^{\infty} \delta(x-a)\, dx$$

$$= f(a)\int_{-\infty}^{\infty} \delta(x-a)\, d(x-a) = f(a).$$

Hence

$$f(x)\, \delta(x-a) = f(a)\, \delta(x-a).$$

N.B. In all these results it would be wrong to cancel terms from each side. For example, in (d), $f(x)$ is not equal to $f(a)$ by cancelling the $\delta(x-a)$ from each side. This is because the results have to be strictly interpreted under an integration sign.

18.10   Show (in some sense) that

$$\frac{d}{dx} H(x) = \delta(x).$$

From the definition of the δ-function it follows that

$$\int_{-\infty}^{x} \delta(x)\,dx = \begin{cases} 0, & x<0 \\ 1, & x>0 \end{cases} = H(x).$$

Hence 'differentiating' both sides we have the required result. This 'differentiation' however is only a formal operation since the function $H(x)$ is not a continuous function. Nor is the δ-function a *proper* function since it is undefined at $x = 0$.

18.11   Solve the differential equation

$$\frac{dy}{dx} + 3y = \delta(x-2)$$

given $y(0) = 1$.

Consider the Laplace transform of $\delta(x-a)$. Then, using the basic results of 18.9, we have

$$L\{\delta(x-a)\} = \int_{0}^{\infty} e^{-px}\,\delta(x-a)\,dx = e^{-pa} \quad \text{for} \quad a>0.$$

Also

$$L\{f(x-a)H(x-a)\} = \int_{0}^{\infty} f(x-a)H(x-a)e^{-px}\,dx$$
$$= \int_{a}^{\infty} f(x-a)e^{-px}\,dx = e^{-ap}\bar{f}(p).$$

Hence the Laplace transform of the equation becomes

$$p\bar{y}(p) - y(0) + 3\bar{y}(p) = e^{-2p}$$

and inserting the initial condition we have

$$\bar{y}(p) = \frac{1}{p+3} + \frac{e^{-2p}}{p+3}.$$

Inversion gives (using the result $L^{-1}\{e^{-ap}\,\bar{f}(p)\} = f(x-a)H(x-a)$ with $a = 2$, $\bar{f}(p) = 1/(p+3)$.

$$y(x) = L^{-1}\{\bar{y}(p)\} = e^{-3x} + H(x-2)e^{-3(x-2)}.$$

18.12   Making the assumption (which can be justified in most cases) that

$$L\left\{\int_0^\infty f(x, u)\, du\right\} = \int_0^\infty L\{f(x, u))\}\, du,$$

where x is a parameter, show that

$$\int_0^\infty \frac{\sin xu}{u}\, du = \frac{\pi}{2}.$$

Let

$$f(x, u) = \frac{\sin xu}{u}.$$

Then

$$L\left\{\int_0^\infty \frac{\sin xu}{u}\, du\right\} = \int_0^\infty L\left\{\frac{\sin xu}{u}\right\}\, du.$$

Now

$$L\left\{\frac{\sin xu}{u}\right\} = \int_0^\infty e^{-px}\frac{\sin xu}{u}\, dx = \frac{1}{p^2+u^2}\quad\text{(by integration by parts)}.$$

Hence

$$\int_0^\infty L\left\{\frac{\sin xu}{u}\right\}\, du = \int_0^\infty \frac{1}{p^2+u^2}\, du = \frac{1}{p}\left(\tan^{-1}\frac{u}{p}\right)_0^\infty = \frac{\pi}{2p}.$$

Therefore

$$L\left\{\int_0^\infty \frac{\sin xu}{u}\, du\right\} = \frac{\pi}{2p}$$

and hence

$$\int_0^\infty \frac{\sin xu}{u}\, du = L^{-1}\left\{\frac{\pi}{2p}\right\} = \frac{\pi}{2}.$$

18.13 The convolution theorem states that under certain conditions

$$L\left\{\int_0^x f(x-x')g(x')\, dx'\right\} = L\left\{\int_0^x g(x-x')f(x')\, dx'\right\} = \bar{f}(p)\bar{g}(p),$$

where $\bar{f}(p) = L\{f(x)\}$ and $\bar{g}(p) = L\{g(x)\}$ (x' is the variable of integration). Show, using the Laplace transform method, that the equation

$$(D^2+4)y = f(x),$$

where $f(x)$ is an arbitrary function, and where $y(0) = 1$, $y'(0) = 1$, becomes

$$\bar{y}(p) = \frac{p+1}{p^2+4} + \frac{\bar{f}(p)}{p^2+4}.$$

Hence, using the convolution theorem, show that

$$y(x) = \cos 2x + \tfrac{1}{2}\sin 2x + \tfrac{1}{2}\int_0^x f(x')\sin 2(x-x')\,dx',$$

where x' is an arbitrary parameter of integration.

Transforming the equation and inserting initial conditions we find

$$p^2\bar{y}(p) - p - 1 + 4\bar{y}(p) = \bar{f}(p)$$

or

$$\bar{y}(p) = \frac{p+1}{p^2+4} + \frac{\bar{f}(p)}{p^2+4} = \frac{p}{p^2+4} + \frac{1}{2}\frac{2}{p^2+4} + \frac{\bar{f}(p)}{p^2+4}.$$

Taking the inverse of the convolution theorem in the form

$$L^{-1}\{\bar{f}(p)\bar{g}(p)\} = \int_0^x g(x-x')f(x')\,dx'$$

we can invert the last term in the equation for $\bar{y}(p)$ by letting $\bar{g}(p) = 1/(p^2+4)$. Then $g(x) = \tfrac{1}{2}\sin 2x$ and the solution is

$$y(x) = L^{-1}\{\bar{y}(p)\} = \cos 2x + \tfrac{1}{2}\sin 2x + \tfrac{1}{2}\int_0^x f(x')\sin 2(x-x')\,dx'.$$

18.14 Use the convolution theorem to show that if

$$y(x) = f(x) + \int_0^x g(x-x')y(x')\,dx'$$

then

$$\bar{y}(p) = \frac{\bar{f}(p)}{1 - \bar{g}(p)}.$$

Hence solve the integral equation

$$y(x) = \sin 3x + \int_0^x \sin(x-x')y(x')\,dx'.$$

Taking Laplace transforms and using the convolution theorem for the last term we have the result

$$\bar{y}(p) = \bar{f}(p) + \bar{g}(p)\bar{y}(p) \quad \text{or} \quad \bar{y}(p) = \frac{\bar{f}(p)}{1 - \bar{g}(p)}.$$

For the integral equation given, $g(x)$ is the sine function. Hence $\bar{g}(p) = \dfrac{1}{p^2+1}$. Also $f(x) = \sin 3x$, so $\bar{f}(p) = \dfrac{3}{p^2+9}$. Hence

$$\bar{y}(p) = \left(\frac{3}{p^2+9}\right)\left(\frac{1}{1-1/(p^2+1)}\right) = \frac{3(p^2+1)}{p^2(p^2+9)}.$$

By partial fractions this may be written as

$$\bar{y}(p) = \frac{1}{3p^2} + \frac{8}{3}\frac{1}{p^2+9},$$

which on inversion gives the result

$$y(x) = \tfrac{1}{3}x + \tfrac{8}{9}\sin 3x.$$

18.15 Evaluate $L\{[e^x]\}$, where $[e^x]$ means the integer part of e^x (see 1.13).

$$L\{[e^x]\} = \int_0^\infty [e^x]e^{-px}\,dx. \text{ Put } e^x = t. \text{ Then}$$

$$L\{[t]\} = \int_1^\infty [t]\frac{dt}{t^{p+1}}.$$

Now $[t] = 1$ for $1 \leqslant t < 2$, $[t] = 2$ for $2 \leqslant t < 3$, and so on. Hence

$$L\{[t]\} = \int_1^2 \frac{dt}{t^{p+1}} + \int_2^3 2\frac{dt}{t^{p+1}} + \cdots$$

$$= \frac{1}{p}\left\{\frac{1}{1^p} + \frac{1}{2^p} + \frac{1}{3^p} + \cdots\right\} = \frac{1}{p}\,\xi(p),$$

where $\xi(p) = \sum_{n=1}^\infty 1/n^p$ is the Riemann zeta function. (A famous conjecture by Riemann is that, in the complex plane, $\xi(z) = 0$ has all its roots on the line $z = \tfrac{1}{2}$.)

---

19. FOURIER SERIES

[Dirichlet conditions, Fourier expansions, summation of series, Fourier sine and cosine series]

19.1 Which of the following functions satisfy the Dirichlet conditions (sufficient conditions for the existence of a Fourier series expansion of the function)?

(a) $f(x) = \dfrac{1}{1-x^2}$ for $-\pi < x < \pi$.

(b) $f(x) = \sin\left(\dfrac{1}{x-1}\right)$ for $-\pi < x < \pi$.

(c) $f(x) = H(x) - H(x-1)$ for $-2 < x < 2$.

Basic result

The Dirichlet conditions require that $f(x)$ be single-valued and bounded in the basic interval (taken here as $-\pi < x < \pi$) and extended to other values of x by the periodicity relation

$$f(x + 2\pi k) = f(x), \quad \text{where } k = \pm 1, \pm 2, \pm 3, \ldots$$

Then if $f(x)$ is continuous in the basic interval except at a *finite* number of finite discontinuities, and has only a *finite* number of maxima and minima, it has a Fourier series which converges to $f(x)$ at all points in $-\pi < x < \pi$ where $f(x)$ is continuous. At a point where $f(x)$ has a finite discontinuity, say $x = x_0$, the Fourier series converges to the value

$$\frac{1}{2} \lim_{\delta \to 0} [f(x_0 + \delta) + f(x_0 - \delta)],$$

which is just the mean of the two limiting values as $x \to x_0$ from each side.

(a) The function $f(x) = 1/(1-x^2)$ does not satisfy the Dirichlet conditions in $-\pi < x < \pi$ since there is an infinite discontinuity at $x = \pm 1$.

(b) The graph of $f(x) = \sin(1/(x-1))$ is shown in Fig. 30. $f(x)$ does

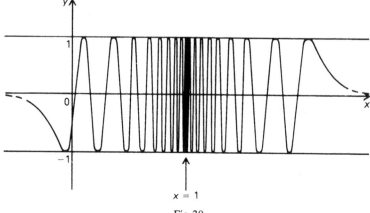

$x = 1$

Fig. 30

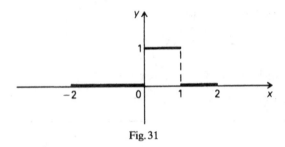

Fig. 31

not satisfy the Dirichlet conditions in $(-\pi, \pi)$ since it has an infinite number of maxima and minima in the neighbourhood of $x = 1$. N.B. $(-\pi, \pi)$ is an alternative notation for $-\pi < x < \pi$.
(c) The graph of $H(x) - H(x - 1)$ is shown in Fig. 31. In the range $(-2, 2)$ the function has two finite discontinuities (at $x = 0, 1$). The Dirichlet conditions are satisfied.

19.2 Show that in the range $-\pi \leqslant x \leqslant \pi$,

$$x^2 = \frac{\pi^2}{3} + 4 \sum_{n=1}^{\infty} \frac{(-1)^n \cos nx}{n^2}.$$

Sketch the graph of the function represented by the series outside the given range, and deduce that

$$\frac{\pi^2}{12} = 1 - \frac{1}{2^2} + \frac{1}{3^2} - \frac{1}{4^2} + \cdots$$

Basic result

A Fourier series in the range $-\pi < x < \pi$ is given by

$$\frac{a_0}{2} + \sum_{n=1}^{\infty} a_n \cos nx + b_n \sin nx,$$

where the Fourier coefficients a_0, a_n and b_n are defined by

$$a_0 = \frac{1}{\pi} \int_{-\pi}^{\pi} f(x)\, dx, \qquad a_n = \frac{1}{\pi} \int_{-\pi}^{\pi} f(x) \cos nx\, dx,$$

$$b_n = \frac{1}{\pi} \int_{-\pi}^{\pi} f(x) \sin nx\, dx.$$

$f(x)$ is extended periodically outside the basic range by the relation

$$f(x + 2\pi k) = f(x), \qquad k = \pm 1, \pm 2, \pm 3, \ldots$$

In this problem $f(x) = x^2$ in $-\pi < x < \pi$ and its periodic extension

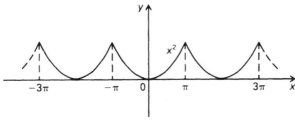

Fig. 32

is shown in Fig. 32.

$$a_0 = \frac{1}{\pi} \int_{-\pi}^{\pi} x^2 \, dx = \frac{2\pi^2}{3},$$

$$a_n = \frac{1}{\pi} \int_{-\pi}^{\pi} x^2 \cos nx \, dx = \frac{2}{\pi} \int_0^{\pi} x^2 \cos nx \, dx$$

$$= \frac{2}{\pi} \left[\left(\frac{x^2 \sin nx}{n} \right)_0^{\pi} - \frac{2}{n} \int_0^{\pi} x \sin nx \, dx \right]$$

$$= -\frac{4}{\pi n} \int_0^{\pi} x \sin nx \, dx = -\frac{4}{\pi n} \left[\left(\frac{-x \cos nx}{n} \right)_0^{\pi} + \frac{1}{n} \int_0^{\pi} \cos nx \, dx \right]$$

$$= \frac{4(-1)^n}{n^2},$$

$b_n = 0$, since the integrand is an odd function. Hence the result. Putting $x = 0$ in the series we have (since, by the Dirichlet conditions, the series converges to the function at this point)

$$\frac{\pi^2}{12} = 1 - \frac{1}{2^2} + \frac{1}{3^2} - \frac{1}{4^2} + \cdots$$

19.3 Find the Fourier expansion of $f(x) = |x|$ in $-\pi < x < \pi$, and hence deduce that

$$\sum_{n=1}^{\infty} \frac{1}{(2n-1)^2} = \frac{\pi^2}{8}.$$

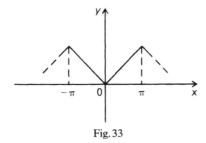

Fig. 33

The graph of the function is shown in Fig. 33. Now

$$a_0 = \frac{1}{\pi} \int_{-\pi}^{\pi} |x| \, dx = \frac{1}{\pi} \int_{-\pi}^{0} (-x) \, dx + \frac{1}{\pi} \int_{0}^{\pi} x \, dx$$

$$= \frac{1}{\pi} \left(\frac{x^2}{2}\right)_0^{-\pi} + \frac{1}{\pi} \left(\frac{x^2}{2}\right)_0^{\pi} = \pi.$$

$$a_n = \frac{1}{\pi} \int_{-\pi}^{\pi} |x| \cos nx \, dx = \frac{1}{\pi} \int_{-\pi}^{0} -x \cos nx \, dx + \frac{1}{\pi} \int_{0}^{\pi} x \cos nx \, dx$$

$$= \frac{1}{\pi} \int_0^{-\pi} x \cos nx \, dx + \frac{1}{\pi} \int_0^{\pi} x \cos nx \, dx$$

$$= \frac{1}{\pi} \left[\left(\frac{x \sin nx}{n}\right)_0^{-\pi} - \frac{1}{n} \int_0^{-\pi} \sin nx \, dx \right]$$

$$\quad + \frac{1}{\pi} \left[\left(\frac{x \sin nx}{n}\right)_0^{\pi} - \frac{1}{n} \int_0^{\pi} \sin nx \, dx \right]$$

$$= \frac{1}{\pi n^2} [(-1)^n - 1] + \frac{1}{\pi n^2} [(-1)^n - 1] = \frac{2}{\pi n^2} [(-1)^n - 1]$$

$$= \begin{cases} 0, & n \text{ even.} \\ -\dfrac{4}{\pi n^2}, & n \text{ odd.} \end{cases}$$

Also $b_n = 0$, since $|x|$ is an even function. Hence

$$|x| = \frac{\pi}{2} + \sum_{n \text{ odd}}^{\infty} \left(-\frac{4}{\pi n^2}\right) \cos nx = \frac{\pi}{2} - \frac{4}{\pi} \left(\frac{\cos x}{1^2} + \frac{\cos 3x}{3^2} + \cdots\right).$$

Putting $x = 0$, the Fourier series converges to the function (by the Dirichlet conditions) and hence

$$0 = \frac{\pi}{2} - \frac{4}{\pi} \left(\frac{1}{1^2} + \frac{1}{3^2} + \frac{1}{5^2} + \cdots\right)$$

or

$$\frac{\pi^2}{8} = \sum_{n=1}^{\infty} \frac{1}{(2n-1)^2}.$$

19.4 Show that in the range $0 \leq x < \pi/2$, the function $f(x)$ defined by

$$f(x) = \begin{cases} 1, & 0 \leq x < \dfrac{\pi}{2}, \\ 0, & x = \dfrac{\pi}{2}, \\ -1, & \dfrac{\pi}{2} < x \leq \pi, \end{cases}$$

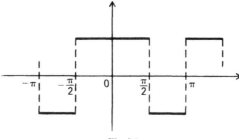

Fig. 34

is represented by the Fourier cosine series

$$f(x) = \frac{4}{\pi} \sum_{n=1}^{\infty} \frac{1}{n} \sin \frac{n\pi}{2} \cos nx.$$

The function is extended into the range $(-\pi, 0)$ as an even function (see Fig. 34). Then the Fourier series of this function in $(-\pi, \pi)$ will be a cosine series, and will represent the given function in the range $(0, \pi)$.

$$a_0 = \frac{2}{\pi} \int_0^{\pi} f(x)\, dx = \frac{2}{\pi} \left[\int_0^{\pi/2} 1\, dx + \int_{\pi/2}^{\pi} (-1)\, dx \right] = 0.$$

$$a_n = \frac{2}{\pi} \int_0^{\pi} f(x)\, dx = \frac{2}{\pi} \int_0^{\pi/2} 1 \cos nx\, dx + \frac{2}{\pi} \int_{\pi/2}^{\pi} (-1) \cos nx\, dx$$

$$= \frac{2}{\pi} \left[\left(\frac{\sin nx}{n} \right)_0^{\pi/2} - \left(\frac{\sin nx}{n} \right)_{\pi/2}^{\pi} \right]$$

$$= \frac{4}{n\pi} \sin \frac{n\pi}{2}.$$

$$b_n = 0.$$

Hence

$$f(x) = \frac{4}{\pi} \sum_{n=1}^{\infty} \frac{1}{n} \sin \frac{n\pi}{2} \cos nx.$$

19.5 Expand the function $\cos x$ in a sine series, valid in $0 < x < \pi$.

The function $\cos x$ is extended into $(-\pi, 0)$ by an odd extension. In this way the full function in $(-\pi, \pi)$ is an odd function and its Fourier series expansion will therefore be a sine series only. This series will represent $\cos x$ in $(0, \pi)$ and its odd extension in $(-\pi, 0)$ (see Fig. 35.).

$$a_0 = a_n = 0$$

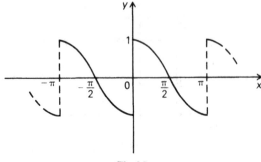

Fig. 35

since the function in $(-\pi, \pi)$ is odd.

$$b_n = \frac{2}{\pi} \int_0^\pi \cos x \sin nx \, dx$$

$$= \frac{2}{\pi} \left[(\sin x \sin nx)_0^\pi - n \int_0^\pi \cos nx \sin x \, dx \right]$$

$$= -\frac{2n}{\pi} \left[\int_0^\pi \cos nx \sin x \, dx \right]$$

$$= -\frac{2n}{\pi} \left[(-\cos x \cos nx)_0^\pi - n \int_0^\pi \cos x \sin nx \, dx \right].$$

Hence

$$\int_0^\pi \cos x \sin nx \, dx = -n[1 + (-1)^n] + n^2 \int_0^\pi \cos x \sin nx \, dx.$$

Hence

$$\int_0^\pi \cos x \sin nx \, dx = \frac{n[1 + (-1)^n]}{n^2 - 1} \quad \text{and} \quad b_n = \frac{2n}{\pi} \frac{1 + (-1)^n}{(n^2 - 1)}.$$

Accordingly

$$\cos x = \frac{2}{\pi} \sum_{n=1}^\infty \frac{[1 + (-1)^n] n \sin nx}{(n^2 - 1)}.$$

At $x = 0$ there is a point of finite discontinuity (the function changing from -1 to $+1$. Hence by the Dirichlet conditions, the series cannot converge to the function but only to the average of the limiting values as x approaches 0. The average is $\frac{1}{2}(1 + (-1)) = 0$ as can be seen from the graph. We check this directly from the series since by putting $x = 0$ in the series we obtain zero, as required.

19.6 Show that the general term in the Fourier half-range series for

$$f(x) = 1 + \frac{x}{l} \qquad (0 < x < l),$$

is

$$\frac{2}{n\pi}(1 - 2\cos n\pi)\sin\frac{n\pi x}{l}.$$

By considering the function represented by this series in the range $l < x < 2l$, and the sum of the series at $x = 3l/2$, prove that

$$\frac{\pi}{4} = 1 - \tfrac{1}{3} + \tfrac{1}{5} - \tfrac{1}{7} + \cdots$$

For the general range $-l < x < l$, the Fourier series takes the form

$$\frac{a_0}{2} + \sum_{n=1}^{\infty}\left(a_n\cos\frac{n\pi x}{l} + b_n\sin\frac{n\pi x}{l}\right),$$

where

$$a_0 = \frac{1}{l}\int_{-l}^{l} f(x)\,\mathrm{d}x, \qquad a_n = \frac{1}{l}\int_{-l}^{l} f(x)\cos\frac{n\pi x}{l}\,\mathrm{d}x,$$

$$b_n = \frac{1}{l}\int_{-l}^{l} f(x)\sin\frac{n\pi x}{l}\,\mathrm{d}x.$$

We now extend the function into $(-l, 0)$ so that the whole function in $(-l, l)$ is an odd function. In this way the function in $(-l, l)$ will be represented by a Fourier sine series, and this expansion will represent the given function in the required range $0 < x < l$ (see Fig. 36.).

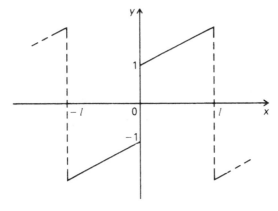

Fig. 36

Now $a_0 = a_n = 0$. In $(-l, 0)$ the extended function has the form $-(1 - x/l)$ (since $f(x) = -f(-x)$ for $f(x)$ to be odd).

$$b_n = \frac{1}{l} \int_{-l}^{0} -\left(1 - \frac{x}{l}\right) \sin \frac{n\pi x}{l} \, dx + \frac{1}{l} \int_{0}^{l} \left(1 + \frac{x}{l}\right) \sin \frac{n\pi x}{l} \, dx$$

$$= \frac{2}{l} \int_{0}^{l} \left(1 + \frac{x}{l}\right) \sin \frac{n\pi x}{l} \, dx$$

$$= \frac{2}{n\pi} \left[\left(-\frac{l}{n\pi} \cos \frac{n\pi x}{l}\right)_{0}^{l} + \frac{1}{l} \int_{0}^{l} x \sin \frac{n\pi x}{l} \, dx \right]$$

$$= \frac{2}{n\pi} [-(\cos n\pi - 1)] + \frac{2}{l^2} \left[\left(-\frac{lx}{n\pi} \cos \frac{n\pi x}{l}\right)_{0}^{l} \right.$$
$$\left. + \left(\frac{l}{n\pi}\right)^2 \left(\sin \frac{n\pi x}{l}\right)_{0}^{l} \right].$$

Hence

$$b_n = \frac{2}{n\pi} [1 - \cos n\pi] - \frac{2}{n\pi} \cos n\pi = \frac{2}{n\pi} (1 - 2 \cos n\pi).$$

The value of $f(x)$ at $x = 3l/2$ is the same as the value at $x = -l/2$ (see Fig. 36.) and is $-3/2$. Inserting these values into the series gives

$$-\tfrac{3}{2} = \frac{2}{\pi} \sum_{n=1}^{\infty} \left(\frac{1 - 2 \cos n\pi}{n}\right) \sin \left(\frac{3n\pi}{2}\right)$$

whence the series for $\pi/4$ follows.

19.7 Show that it is not valid to differentiate the Fourier series

$$x = 2(\sin x - \tfrac{1}{2} \sin 2x + \tfrac{1}{3} \sin 3x - \cdots),$$

where $-\pi < x < \pi$.

Differentiating the left-hand side gives unity. Differentiating the right-hand side gives the series

$$2(\cos x - \cos 2x + \cos 3x - \cdots)$$

The nth term of this series is $a_n = (-1)^{n-1} \cos nx$ and the necessary condition for convergence $\lim_{n \to \infty} a_n = 0$ is clearly not satisfied. The series is therefore divergent and consequently cannot be equal to unity. Hence differentiation of the given Fourier series is not valid.

20. VECTOR CALCULUS

[Vector functions, gradient fields, grad, div and curl operators, Green's theorem in the plane, Stokes' theorem, divergence theorem]

20.1   Given that $\mathbf{A} = e^{uv}\mathbf{i} + (2u - v)\mathbf{j} + (v \sin u)\mathbf{k}$, calculate

(a) $\dfrac{\partial \mathbf{A}}{\partial u}$,   (b) $\dfrac{\partial^2 \mathbf{A}}{\partial u^2}$,   (c) $\left(\dfrac{\partial^2}{\partial u^2} + \dfrac{\partial^2}{\partial v^2}\right)\mathbf{A}$

when $u = \pi/2$ and $v = 0$.

$$\frac{\partial \mathbf{A}}{\partial u} = v e^{uv}\mathbf{i} + 2\mathbf{j} + (v \cos u)\mathbf{k},$$

$$\frac{\partial^2 \mathbf{A}}{\partial u^2} = v^2 e^{uv}\mathbf{i} - (v \sin u)\mathbf{k}.$$

$$\frac{\partial \mathbf{A}}{\partial v} = u e^{uv}\mathbf{i} - \mathbf{j} + (\sin u)\mathbf{k},$$

$$\frac{\partial^2 \mathbf{A}}{\partial v^2} = u^2 e^{uv}\mathbf{i}.$$

Hence

$$\left(\frac{\partial^2}{\partial u^2} + \frac{\partial^2}{\partial v^2}\right)\mathbf{A} = (u^2 + v^2)e^{uv}\mathbf{i} - (v \sin u)\mathbf{k}.$$

At $u = \pi/2$, $v = 0$, the result is therefore $(\pi^2/4)\mathbf{i}$.

N.B. The operation $\dfrac{\partial^2}{\partial u^2} + \dfrac{\partial^2}{\partial v^2}$ is frequently denoted by ∇^2 (see 20.9).

20.2   Obtain gradients of the following scalar fields:

(a) x,   (b) $x^3 + y^3 + z^3$,   (c) r^n,

(d) $\mathbf{a} \cdot \mathbf{r}$ where $\mathbf{a}$ is a constant vector,

(e) $\mathbf{r} \cdot \nabla(x + y + z)$,   (f) $\nabla f(r)$.

In all these, the notation is

$$\mathbf{r} = \mathbf{i}x + \mathbf{j}y + \mathbf{k}z, \qquad |\mathbf{r}| = (\mathbf{r} \cdot \mathbf{r})^{1/2} = r.$$

Basic result

The ∇ operator is defined (in rectangular Cartesian coordinates) as

$$\mathbf{i}\frac{\partial}{\partial x} + \mathbf{j}\frac{\partial}{\partial y} + \mathbf{k}\frac{\partial}{\partial z},$$

and if ϕ is any scalar field then the gradient of ϕ, denoted by grad ϕ, is

$$\nabla\phi = \mathbf{i}\frac{\partial\phi}{\partial x} + \mathbf{j}\frac{\partial\phi}{\partial y} + \mathbf{k}\frac{\partial\phi}{\partial z}.$$

(a) $\nabla x = \left(\mathbf{i}\dfrac{\partial}{\partial x} + \mathbf{j}\dfrac{\partial}{\partial y} + \mathbf{k}\dfrac{\partial}{\partial z}\right)x = \mathbf{i}.$

(b) $\nabla(x^3 + y^3 + z^3) = 3x^2\mathbf{i} + 3y^2\mathbf{j} + 3z^2\mathbf{k}.$

(c) $r^2 = x^2 + y^2 + z^2, \qquad r^n = (x^2 + y^2 + z^2)^{n/2}.$

$$\nabla r^n = \frac{n}{2} \cdot 2x(x^2 + y^2 + z^2)^{(n/2)-1}\mathbf{i} + \frac{n}{2} \cdot 2y(x^2 + y^2 + z^2)^{(n/2)-1}\mathbf{j}$$

$$+ \frac{n}{2} \cdot 2z(x^2 + y^2 + z^2)^{(n/2)-1}\mathbf{k}.$$

Hence $\nabla r^n = nr^{n-2}\mathbf{r}.$

(d) $\nabla(\mathbf{a} \cdot \mathbf{r}) = \nabla(a_x x + a_y y + a_z z) = \mathbf{i}a_x + \mathbf{j}a_y + \mathbf{k}a_z = \mathbf{a}.$

(e) $\nabla[\mathbf{r} \cdot \nabla(x + y + z)] = \nabla[\mathbf{r} \cdot (\mathbf{i} + \mathbf{j} + \mathbf{k})] = \nabla(x + y + z) = \mathbf{i} + \mathbf{j} + \mathbf{k}.$

(f) $\nabla f(r) = \mathbf{i}\dfrac{x}{r}\dfrac{\partial f}{\partial r} + \mathbf{j}\dfrac{y}{r}\dfrac{\partial f}{\partial r} + \mathbf{k}\dfrac{z}{r}\dfrac{\partial f}{\partial r} = \hat{\mathbf{r}}\dfrac{\partial f}{\partial r},$

where $\hat{\mathbf{r}}$ is the *unit* normal in the direction of $\mathbf{r}$.

20.3 Find the condition that the plane $ax + by + cz + d = 0$ should touch the surface $Ax^2 + By^2 + Cz^2 = 1$ (a, A, b, B, c, C, d are constants).

The normal to plane is in the direction of $\nabla\phi$, where $\phi = ax + by + cz + d$, i.e. in the direction $a\mathbf{i} + b\mathbf{j} + c\mathbf{k}$.
The normal to curved surface is in the direction $\nabla\theta$, where $\theta = Ax^2 + By^2 + Cz^2 - 1$, i.e. in the direction $2Ax\mathbf{i} + 2By\mathbf{j} + 2Cz\mathbf{k}$.

Hence for the plane to touch the surface tangentially these two normals must be in the same direction. Therefore at the point of contact

$$2Ax\mathbf{i} + 2By\mathbf{j} + 2Cz\mathbf{k} = \lambda(a\mathbf{i} + b\mathbf{j} + c\mathbf{k}),$$

where λ is some constant. Comparing corresponding terms, the point of contact is given by

$$x = \frac{\lambda a}{2A}, \qquad y = \frac{\lambda b}{2B}, \qquad z = \frac{\lambda c}{2C}.$$

Inserting these expressions for x, y, z into the equations defining the plane and the surface we have

$$A\left(\frac{\lambda a}{2A}\right)^2 + B\left(\frac{\lambda b}{2B}\right)^2 + C\left(\frac{\lambda c}{2C}\right)^2 = 1, \qquad a\frac{\lambda a}{2A} + b\frac{\lambda b}{2B} + c\frac{\lambda c}{2C} + d = 0.$$

Hence

$$\frac{a^2}{A} + \frac{b^2}{B} + \frac{c^2}{C} = \frac{4}{\lambda^2}$$

and

$$\frac{a^2}{A} + \frac{b^2}{B} + \frac{c^2}{C} + \frac{2d}{\lambda} = 0, \quad \text{whence} \quad \lambda = -\frac{2}{d}.$$

Therefore

$$\frac{a^2}{A} + \frac{b^2}{B} + \frac{c^2}{C} = d^2$$

is the required condition.

20.4 Prove that the vector field

$$\mathbf{F} = (3x^2 + yz)\mathbf{i} + z(x-1)\mathbf{j} + y(x-1)\mathbf{k}$$

is conservative by finding a scalar field ϕ such that $\mathbf{F} = \text{grad } \phi$. Hence evaluate $\int \mathbf{F} \cdot d\mathbf{r}$ along the curve defined by

$$\mathbf{r} = t\mathbf{i} - t^2\mathbf{j} + t^3\mathbf{k}$$

from $(0, 0, 0)$ to $(1, -1, 1)$.

Now

$$\mathbf{F} = \mathbf{i}\frac{\partial \phi}{\partial x} + \mathbf{j}\frac{\partial \phi}{\partial y} + \mathbf{k}\frac{\partial \phi}{\partial z},$$

where

$$\frac{\partial \phi}{\partial x} = 3x^2 + yz, \qquad \frac{\partial \phi}{\partial y} = z(x-1), \qquad \frac{\partial \phi}{\partial z} = y(x-1).$$

Therefore integrating each expression

$$\phi = x^3 + xyz + \lambda(y, z), \quad \phi = zxy - zy + \mu(x, z), \quad \phi = xyz - yz + \nu(x, y),$$

where λ, μ, ν are arbitrary functions of integration. Comparing the three forms for ϕ we find $\mu = \nu = x^3$, $\lambda = -zy$, and hence

$$\phi = x^3 + xyz - zy.$$

Accordingly (since the end points correspond to $t = 0$ and $t = 1$)

$$\int_{t=0}^{1} \mathbf{F} \cdot d\mathbf{r} = \int_{0}^{1} \nabla\phi \cdot d\mathbf{r} = [\phi]_0^1 = 1.$$

20.5 Obtain divergences of the following vector functions:

(a) $y\mathbf{i} + z\mathbf{j} + x\mathbf{k}$,   (b) $\mathbf{r}$,   (c) $xyz(\mathbf{i} + \mathbf{j} + \mathbf{k})$,   (d) $\mathbf{r}/r^3$.

Basic result

$$\text{div } \mathbf{A} = \nabla \cdot \mathbf{A} = \mathbf{i}\frac{\partial A_x}{\partial x} + \mathbf{j}\frac{\partial A_y}{\partial y} + \mathbf{k}\frac{\partial A_z}{\partial z} \qquad \text{in Cartesian coordinates.}$$

(a) $\nabla \cdot (y\mathbf{i} + z\mathbf{j} + x\mathbf{k}) = \mathbf{i}\dfrac{\partial y}{\partial x} + \mathbf{j}\dfrac{\partial z}{\partial y} + \mathbf{k}\dfrac{\partial x}{\partial z} = 0.$

(b) $\nabla \cdot (\mathbf{i}x + \mathbf{j}y + \mathbf{k}z) = 3.$

(c) $\nabla \cdot (xyz(\mathbf{i} + \mathbf{j} + \mathbf{k})) = yz + xz + xy.$

(d) $\nabla \cdot \left(\dfrac{\mathbf{r}}{r^3}\right) = \dfrac{\partial}{\partial x}\left(\dfrac{x}{r^3}\right) + \dfrac{\partial}{\partial y}\left(\dfrac{y}{r^3}\right) + \dfrac{\partial}{\partial z}\left(\dfrac{z}{r^3}\right)$

$$= \frac{1}{r^3} - \frac{3x}{r^4}\frac{x}{r} + \frac{1}{r^3} - \frac{3y}{r^4}\frac{y}{r} + \frac{1}{r^3} - \frac{3z}{r^4}\frac{z}{r}$$

$$= \frac{3}{r^3} - \frac{3}{r^5}(x^2 + y^2 + z^2) = 0.$$

20.6 Obtain the curls of the following vectors:
(a) $x\mathbf{i}$, (b) $\mathbf{r}$, (c) $f(r)\mathbf{r}$, where $f(r)$ is an arbitrary differentiable function, (d) $(x\mathbf{i} - y\mathbf{j})/(x + y)$.

Basic result

$$\text{curl } \mathbf{A} = \nabla \wedge \mathbf{A} = \mathbf{i}\left(\frac{\partial A_z}{\partial y} - \frac{\partial A_y}{\partial z}\right) + \mathbf{j}\left(\frac{\partial A_x}{\partial z} - \frac{\partial A_z}{\partial x}\right) + \mathbf{k}\left(\frac{\partial A_y}{\partial x} - \frac{\partial A_x}{\partial y}\right)$$

in Cartesian coordinates.

(a) $\mathbf{\nabla} \wedge (x\mathbf{i}) = \left(\mathbf{i} \dfrac{\partial}{\partial x} + \mathbf{j} \dfrac{\partial}{\partial y} + \mathbf{k} \dfrac{\partial}{\partial z} \right) \wedge (x\mathbf{i}) = 0 - \mathbf{k} \dfrac{\partial x}{\partial y} + \mathbf{j} \dfrac{\partial x}{\partial z} = 0$

(using $\mathbf{i} \wedge \mathbf{j} = \mathbf{k}$ etc. – see 13.2).

(b) $\mathbf{\nabla} \wedge (\mathbf{r}) = \left(\mathbf{i} \dfrac{\partial}{\partial x} + \mathbf{j} \dfrac{\partial}{\partial y} + \mathbf{k} \dfrac{\partial}{\partial z} \right) \wedge (\mathbf{i}x + \mathbf{j}y + \mathbf{k}z)$

$= 0$ (again using the vector products of $\mathbf{i}$, $\mathbf{j}$, $\mathbf{k}$).

(c) Here we use the identity

$$\mathbf{\nabla} \wedge (\phi \mathbf{A}) = \phi \mathbf{\nabla} \wedge \mathbf{A} + (\mathbf{\nabla} \phi) \wedge \mathbf{A},$$

where ϕ is some scalar function. With $\phi = f(r)$, $\mathbf{A} = \mathbf{r}$ we have

$$\mathbf{\nabla} \wedge (f(r)\mathbf{r}) = f(r) \mathbf{\nabla} \wedge \mathbf{r} + (\mathbf{\nabla} f(r)) \wedge \mathbf{r} = \hat{\mathbf{r}} \dfrac{\partial f}{\partial r} \wedge \mathbf{r} = \dfrac{\partial f}{\partial r} \hat{\mathbf{r}} \wedge \mathbf{r} = 0,$$

since $\mathbf{\nabla} f(r) = \hat{\mathbf{r}} (\partial f / \partial r)$ (see 20.2(f)), and $\mathbf{\nabla} \wedge \mathbf{r} = 0$.

(d) $\mathbf{\nabla} \wedge \left(\dfrac{x\mathbf{i} - y\mathbf{j}}{x + y} \right) = \left(\mathbf{i} \dfrac{\partial}{\partial x} + \mathbf{j} \dfrac{\partial}{\partial y} + \mathbf{k} \dfrac{\partial}{\partial z} \right) \wedge \left(\dfrac{x\mathbf{i}}{x + y} - \dfrac{y\mathbf{j}}{x + y} \right)$

$= -\mathbf{k} \dfrac{\partial}{\partial y} \left(\dfrac{x}{x + y} \right) - \mathbf{k} \dfrac{\partial}{\partial x} \left(\dfrac{y}{x + y} \right)$

$= \mathbf{k} \dfrac{x}{(x + y)^2} + \mathbf{k} \dfrac{y}{(x + y)^2} = \mathbf{k} \dfrac{1}{x + y}.$

20.7   Show that if $\mathbf{s} = \mathbf{r}(\mathbf{a} \cdot \mathbf{r})$, where $\mathbf{a}$ is a constant vector, then

div $\mathbf{s} = 4(\mathbf{a} \cdot \mathbf{r})$,   curl $\mathbf{s} = \mathbf{a} \wedge \mathbf{r}$,   and   curl curl $\mathbf{s} = $ curl $(\mathbf{a} \wedge \mathbf{r}) = 2\mathbf{a}$.

We need the basic identity

$$\mathbf{\nabla} \cdot (\phi \mathbf{A}) = (\mathbf{\nabla} \phi) \cdot \mathbf{A} + \phi \mathbf{\nabla} \cdot \mathbf{A},$$

where ϕ is a scalar field. Then with $\phi = \mathbf{a} \cdot \mathbf{r}$, $\mathbf{A} = \mathbf{r}$,

$$\mathbf{\nabla} \cdot (\mathbf{r}(\mathbf{a} \cdot \mathbf{r})) = \mathbf{\nabla} (\mathbf{a} \cdot \mathbf{r}) \cdot \mathbf{r} + (\mathbf{a} \cdot \mathbf{r}) \mathbf{\nabla} \cdot \mathbf{r}.$$

From 20.2(d) $\mathbf{\nabla}(\mathbf{a} \cdot \mathbf{r}) = \mathbf{a}$, and from 20.5(b) $\mathbf{\nabla} \cdot \mathbf{r} = 3$. Hence

$$\mathbf{\nabla} \cdot (\mathbf{r}(\mathbf{a} \cdot \mathbf{r})) = 4(\mathbf{a} \cdot \mathbf{r}).$$

To evaluate curl $\mathbf{s}$ we again need the identity given in 20.6, namely

$$\mathbf{\nabla} \wedge (\phi \mathbf{A}) = \phi \mathbf{\nabla} \wedge \mathbf{A} + (\mathbf{\nabla} \phi) \wedge \mathbf{A}.$$

Hence taking $\phi = \mathbf{a} \cdot \mathbf{r}$, $\mathbf{A} = \mathbf{r}$, we have

$$\mathbf{\nabla} \wedge (\mathbf{r}(\mathbf{a} \cdot \mathbf{r})) = (\mathbf{a} \cdot \mathbf{r}) \mathbf{\nabla} \wedge \mathbf{r} + \mathbf{\nabla}(\mathbf{a} \cdot \mathbf{r}) \wedge \mathbf{r} = \mathbf{a} \wedge \mathbf{r},   \text{as required.}$$

For curl $(\mathbf{a} \wedge \mathbf{r})$ we use the basic identity

$$\nabla \wedge (\mathbf{A} \wedge \mathbf{B}) = (\mathbf{B} \cdot \nabla)\mathbf{A} - (\mathbf{A} \cdot \nabla)\mathbf{B} + \mathbf{A}(\nabla \cdot \mathbf{B}) - \mathbf{B}(\nabla \cdot \mathbf{A}),$$

where $\mathbf{A}$ and $\mathbf{B}$ are two arbitrary vector fields.
Then

$$\text{curl } (\mathbf{a} \wedge \mathbf{r}) = (\mathbf{r} \cdot \nabla)\mathbf{a} - (\mathbf{a} \cdot \nabla)\mathbf{r} + \mathbf{a}(\nabla \cdot \mathbf{r}) - \mathbf{r}(\nabla \cdot \mathbf{a})$$
$$= 0 - \mathbf{a} + 3\mathbf{a} - 0 = 2\mathbf{a}.$$

20.8   Show that if $\mathbf{a}$ is a constant unit vector, then

$$\mathbf{a} \cdot [\nabla(\mathbf{v} \cdot \mathbf{a}) - \nabla \wedge (\mathbf{v} \wedge \mathbf{a})] = \nabla \cdot \mathbf{v},$$

where $\mathbf{v}$ is an arbitrary vector.

Using the basic identities

$$\nabla(\mathbf{A} \cdot \mathbf{B}) = (\mathbf{A} \cdot \nabla)\mathbf{B} + (\mathbf{B} \cdot \nabla)\mathbf{A} + \mathbf{A} \wedge \nabla \wedge \mathbf{B} + \mathbf{B} \wedge \nabla \wedge \mathbf{A},$$
$$\nabla \wedge (\mathbf{A} \wedge \mathbf{B}) = (\mathbf{B} \cdot \nabla)\mathbf{A} - (\mathbf{A} \cdot \nabla)\mathbf{B} + \mathbf{A}(\nabla \cdot \mathbf{B}) - \mathbf{B}(\nabla \cdot \mathbf{A}),$$

we have

$$\nabla(\mathbf{v} \cdot \mathbf{a}) - \nabla \wedge (\mathbf{v} \wedge \mathbf{a}) = \mathbf{a} \wedge \nabla \wedge \mathbf{v} + \mathbf{a}(\nabla \cdot \mathbf{v}),$$

the terms $(\mathbf{v} \cdot \nabla)\mathbf{a}$ and $\nabla \cdot \mathbf{a}$ being zero since $\mathbf{a}$ is a constant vector. Hence

$$\mathbf{a} \cdot [\nabla(\mathbf{v} \cdot \mathbf{a}) - \nabla \wedge (\mathbf{v} \wedge \mathbf{a})] = \mathbf{a} \cdot [\mathbf{a} \wedge \nabla \wedge \mathbf{v} + \mathbf{a}(\nabla \cdot \mathbf{v})] = \nabla \cdot \mathbf{v}$$

since $\mathbf{a}$ is a *unit* vector. N.B. $\mathbf{a} \cdot \mathbf{a} \wedge (\nabla \wedge \mathbf{v}) = 0$ since it is a scalar triple product with the form $\mathbf{a} \cdot (\mathbf{a} \wedge \mathbf{b})$ which is identically zero.

20.9   (a) Show that $\nabla^2 r^n = n(n + 1)r^{n-2}$, where $\nabla^2 = \nabla \cdot \nabla$.

(b) Show that if curl $\mathbf{a} = \mathbf{b}$ and curl $\mathbf{b} = \mathbf{a}$, then $\nabla^2 \mathbf{b} + \mathbf{b} = 0$, where the vectors are defined in terms of rectangular Cartesian coordinates.

(a) Now $\nabla r^n = nr^{n-2}\mathbf{r}$ (see 20.2(c)). Hence

$$\nabla^2 r^n = \nabla \cdot \nabla r^n = \nabla \cdot [nr^{n-2}\mathbf{r}] = n[\nabla(r^{n-2}) \cdot \mathbf{r} + r^{n-2} \nabla \cdot \mathbf{r}],$$

using the basic identity of 20.7 with $\phi = r^{n-2}$, $\mathbf{A} = \mathbf{r}$. Hence

$$\nabla^2 r^n = n[(n - 2)r^{n-4}\mathbf{r} \cdot \mathbf{r} + 3r^{n-2}] = n(n + 1)r^{n-2}.$$

(b) Here we use the basic result (valid only in rectangular Cartesian coordinates)

$$\text{curl curl } \mathbf{b} = \text{grad div } \mathbf{b} - \nabla^2 \mathbf{b},$$

where

$$\nabla^2 = \frac{\partial^2}{\partial x^2} + \frac{\partial^2}{\partial y^2} + \frac{\partial^2}{\partial z^2}.$$

Now   curl $\mathbf{b} = \mathbf{a}$ and   curl $\mathbf{a} = \mathbf{b}$,   so $\mathbf{b} = \operatorname{grad} \operatorname{div} \mathbf{b} - \nabla^2 \mathbf{b}$.   But div curl $\mathbf{a} = 0$ identically. Hence div $\mathbf{b} = 0$, and therefore

$$\nabla^2 \mathbf{b} + \mathbf{b} = 0.$$

20.10   By putting $P = -f \dfrac{\partial g}{\partial y}$ and $Q = f \dfrac{\partial g}{\partial x}$ in Green's theorem in the plane, where f and g are functions of x and y, show that

$$\iint_R (f \nabla^2 g + \nabla f \cdot \nabla g)\, dA = \oint_C f \nabla g \cdot \mathbf{n}\, ds,$$

where R is the region enclosed by the simple closed curve C, ds is the element of arc length of C, and $\mathbf{n} = (dy/ds, -dx/ds)$ is the outward normal to the curve at the point $(x(s), y(s))$.   (L.U.)

Use Green's theorem in the plane (see 15.12)

$$\iint_R \left(\frac{\partial Q}{\partial x} - \frac{\partial P}{\partial y} \right) dA = \oint_C P\, dx + Q\, dy.$$

Now

$$\frac{\partial Q}{\partial x} = \frac{\partial}{\partial x}\left(f \frac{\partial g}{\partial x} \right) = \frac{\partial f}{\partial x}\frac{\partial g}{\partial x} + f \frac{\partial^2 g}{\partial x^2}.$$

$$\frac{\partial P}{\partial y} = \frac{\partial}{\partial y}\left(-f \frac{\partial g}{\partial y} \right) = -\frac{\partial f}{\partial y}\frac{\partial g}{\partial y} - f \frac{\partial^2 g}{\partial y^2}.$$

Therefore

$$\frac{\partial Q}{\partial x} - \frac{\partial P}{\partial y} = f\left(\frac{\partial^2 g}{\partial x^2} + \frac{\partial^2 g}{\partial y^2} \right) + \frac{\partial f}{\partial x}\frac{\partial g}{\partial x} + \frac{\partial f}{\partial y}\frac{\partial g}{\partial y} = f \nabla^2 g + \nabla f \cdot \nabla g.$$

Hence

$$\iint_R [f \nabla^2 g + \nabla f \cdot \nabla g]\, dA = \oint_C \left(-f \frac{\partial g}{\partial y} dx + f \frac{\partial g}{\partial x} dy \right)$$

$$= \oint_C f \nabla g \cdot \mathbf{n}\, ds,$$

since

$$\nabla g = \mathbf{i} \frac{\partial g}{\partial x} + \mathbf{j} \frac{\partial g}{\partial y}, \qquad \mathbf{n} = \mathbf{i} \frac{dy}{ds} - \mathbf{j} \frac{dx}{ds}$$

and

$$\nabla g \cdot \mathbf{n} = \frac{\partial g}{\partial x}\frac{dy}{ds} - \frac{\partial g}{\partial y}\frac{dx}{ds}.$$

20.11   Show that if

$$\mathbf{F} = \text{grad } \phi \wedge \text{grad } \psi,$$
$$\mathbf{A} = \tfrac{1}{2}(\phi \text{ grad } \psi - \psi \text{ grad } \phi),$$

then $\mathbf{F} = \text{curl } \mathbf{A}$.

Using the basic identity 20.6(c) for the curl of the product of a scalar and a vector we have

$$\text{curl } \mathbf{A} = \tfrac{1}{2}[\phi \text{ curl grad } \psi + \text{grad } \phi \wedge \text{grad } \psi$$
$$- \psi \text{ curl grad } \phi - \text{grad } \psi \wedge \text{grad } \phi]$$
$$= \text{grad } \phi \wedge \text{grad } \psi,$$

since curl grad $\phi = 0$ identically. Hence

$$\text{curl } \mathbf{A} = \mathbf{F}.$$

20.12   Use the divergence theorem (see 20.13)

$$\iiint_V \text{div } \mathbf{A} \, dV = \iint_S \mathbf{A} \cdot d\mathbf{S}$$

to show that

$$\iiint_V F^2 \, dV = \iint_S \phi \mathbf{F} \cdot d\mathbf{S}$$

if $\mathbf{F} = \nabla\phi$, $\nabla \cdot \mathbf{F} = 0$ and V is the volume enclosed by S.

Using the basic result $\nabla \cdot (\phi\mathbf{A}) = (\nabla\phi) \cdot \mathbf{A} + \phi \nabla \cdot \mathbf{A}$, and writing $\mathbf{A} = \mathbf{F}$, we have

$$\nabla \cdot (\phi\mathbf{F}) = \nabla \cdot (\phi \nabla\phi) = \nabla\phi \cdot \nabla\phi + \phi \nabla^2\phi.$$

But $\nabla \cdot \mathbf{F} = 0$ so $\nabla^2\phi = 0$, whence

$$\nabla \cdot (\phi\mathbf{F}) = |\nabla\phi|^2 = F^2.$$

Hence by the divergence theorem

$$\iiint_V \nabla \cdot (\phi\mathbf{F}) \, dV = \iiint_V F^2 \, dV = \iint_S \phi\mathbf{F} \cdot d\mathbf{S}.$$

20.13   Given that $\mathbf{F} = ax\mathbf{i} + by\mathbf{j} + cz\mathbf{k}$ where a, b, c are constants, show, using the divergence theorem, that

$$\iint_S \mathbf{F} \cdot d\mathbf{S} = \frac{4\pi}{3}(a + b + c),$$

where S is a unit sphere centred at the origin.

Basic result

The divergence theorem states

$$\iiint_V \text{div}\,\mathbf{F}\,dV = \iint_S \mathbf{F}\cdot d\mathbf{S},$$

$d\mathbf{S} = \hat{\mathbf{n}}\,dS$, where $\hat{\mathbf{n}}$ is the outward unit normal to dS and S is the bounding surface of the volume V.

$\text{div}\,F = \boldsymbol{\nabla}\cdot\mathbf{F} = a + b + c.$

Therefore

$$\iiint_V (a+b+c)\,dV = \iint_S \mathbf{F}\cdot d\mathbf{S} = \tfrac{4}{3}\pi(a+b+c),$$

since

$$\iiint_V dV = \tfrac{4}{3}\pi \quad (V \text{ being a sphere of unit radius}).$$

20.14   Verify Stokes' theorem for the vector field

$$\mathbf{F} = (2x - y)\mathbf{i} - yz^2\mathbf{j} - y^2z\mathbf{k},$$

where S is the upper half-surface of the sphere $x^2 + y^2 + z^2 = 1$, and C is its bounding curve in the xy-plane (see Fig. 37).

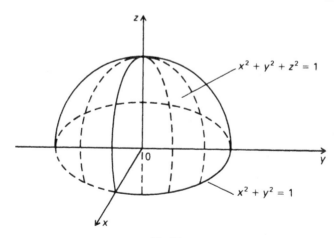

Fig. 37

Basic result

Stokes' theorem states that

$$\iint_S \text{curl } \mathbf{F} \cdot d\mathbf{S} = \oint_C \mathbf{F} \cdot d\mathbf{r},$$

where $d\mathbf{S}$ is the element of vector area of the surface S and $d\mathbf{r}$ is an element of the curve C which bounds S. On $C : x^2 + y^2 = 1$, $z = 0$. Hence

$$\oint_C \mathbf{F} \cdot d\mathbf{r} = \oint_C [(2x - y)\, dx - yz^2\, dy - y^2 z\, dz].$$

But $dz = 0$ since $z = 0$. Let $x = \cos t$, $y = \sin t$. Hence

$$\oint_C \mathbf{F} \cdot d\mathbf{r} = \int_0^{2\pi} (2\cos t - \sin t)(-\sin t)\, dt = \pi.$$

Now

$$\text{curl } \mathbf{F} = \begin{vmatrix} \mathbf{i} & \mathbf{j} & \mathbf{k} \\ \dfrac{\partial}{\partial x} & \dfrac{\partial}{\partial y} & \dfrac{\partial}{\partial z} \\ 2x - y & -yz^2 & -y^2 z \end{vmatrix} = \mathbf{k}.$$

Therefore

$$\iint_S \text{curl } \mathbf{F} \cdot d\mathbf{S} = \iint_S \mathbf{k} \cdot \hat{\mathbf{n}}\, dS,$$

where $\hat{\mathbf{n}}$ is the unit normal to dS such that $d\mathbf{S} = \hat{\mathbf{n}}\, dS$. This is just the projection of the area of S on the xy-plane and is therefore equal to

$$\iint dx\, dy$$

evaluated over the circle $x^2 + y^2 = 1$. Since the area of the circle is π, we have

$$\iint_S \text{curl } \mathbf{F} \cdot d\mathbf{S} = \pi = \oint_C \mathbf{F} \cdot d\mathbf{r},$$

and Stokes' theorem is verified.

20.15 Apply Stokes' theorem to the vector $\mathbf{c}\phi$, where $\mathbf{c}$ is a constant vector and ϕ is a scalar field, to show that

$$\iint_S d\mathbf{S} \wedge \nabla\phi = \oint_C \phi\, d\mathbf{r}.$$

Using the basic result $\nabla \wedge (\phi \mathbf{c}) = \phi \nabla \wedge \mathbf{c} + (\nabla \phi) \wedge \mathbf{c}$ (see 20.6(c)), we have, since $\mathbf{c}$ is a constant,

$$\nabla \wedge (\phi \mathbf{c}) = (\nabla \phi) \wedge \mathbf{c}.$$

Therefore

$$\iint_S \nabla \wedge (\phi \mathbf{c}) \cdot d\mathbf{S} = \iint_S (\nabla \phi \wedge \mathbf{c}) \cdot d\mathbf{S} = \mathbf{c} \cdot \iint_S d\mathbf{S} \wedge \nabla \phi$$

(using the cyclic property of the scalar triple product). By Stokes' theorem

$$\iint_S \nabla \wedge (\phi \mathbf{c}) \cdot d\mathbf{S} = \oint_C \phi \mathbf{c} \cdot d\mathbf{r} = \mathbf{c} \cdot \oint_C \phi \, d\mathbf{r}.$$

Hence it follows that

$$\mathbf{c} \cdot \iint_S d\mathbf{S} \wedge \nabla \phi = \mathbf{c} \cdot \oint_C \phi \, d\mathbf{r}$$

or

$$\iint_S d\mathbf{S} \wedge \nabla \phi = \oint_C \phi \, d\mathbf{r}.$$

20.16 Show that the surface integral of a vector field over a region S of a plane (whose normal $\mathbf{n}$ is not parallel to the Oz axis) may be written as

$$\iint_S \mathbf{F} \cdot d\mathbf{S} = \iint_R \frac{\mathbf{F} \cdot \mathbf{n}}{|\mathbf{n} \cdot \mathbf{k}|} \, dx \, dy,$$

where R is the projection of S onto the xy-plane, and $\mathbf{k}$ is the unit vector in the Oz direction (see Fig. 38).

Evaluate by direct integration the *outward* flux of $\mathbf{F} = (x-2)\mathbf{i} + (x+3y)\mathbf{j} + 2z\mathbf{k}$ over the closed surface of the tetrahedron formed by the planes $x = 0$, $y = 0$, $z = 0$ and $2x + 2y + z = 2$. Verify the result using the divergence theorem.

From Fig. 38, $\mathbf{n} \, dS = d\mathbf{S}$,

$$dx \, dy = dS \, |\cos \theta| = |\mathbf{n} \cdot \mathbf{k}| \, dS.$$

Hence

$$dS = \mathbf{n} \frac{dx \, dy}{|\mathbf{n} \cdot \mathbf{k}|} \quad \text{so} \quad \iint_S \mathbf{F} \cdot d\mathbf{S} = \iint_R \frac{\mathbf{F} \cdot \mathbf{n}}{|\mathbf{n} \cdot \mathbf{k}|} \, dx \, dy.$$

Now consider the tetrahedron shown in Fig. 39, where $A = (1, 0, 0)$,

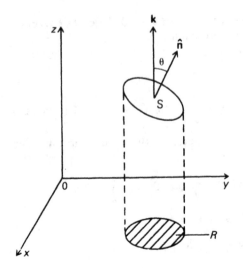

Fig. 38

$B = (0, 1, 0)$, $C = (0, 0, 2)$, AB is the line $x + y = 1$, BC is the line $2y + z = 2$ and AC is the line $2x + z = 2$.

We now have to evaluate $\iint_S \mathbf{F} \cdot d\mathbf{S}$ for each of the four surfaces of the tetrahedron (taking the normal to each surface in the outward sense) and sum the results.

Over AOB, for which $z = 0$,

$$\iint_S \mathbf{F} \cdot d\mathbf{S} = \iint_{z=0} 2\mathbf{k}z \cdot d\mathbf{S} = 0.$$

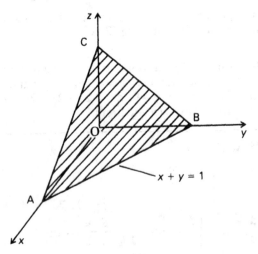

Fig. 39

Over BOC, for which $x = 0$,

$$\iint_S \mathbf{F} \cdot d\mathbf{S} = \iint (x-2)\mathbf{i} \cdot d\mathbf{S} = 2 \iint dz \, dy = 2,$$

since outward normal is in the $-\mathbf{i}$ direction.
Over AOC, for which $y = 0$,

$$\iint_S \mathbf{F} \cdot d\mathbf{S} = \iint_{y=0} (x+3y)\mathbf{j} \cdot d\mathbf{S} = -\int_{x=0}^{1} \int_{z=0}^{2(1-x)} x \, dx \, dz$$

$$= -2 \int_0^1 x(1-x) \, dx = -\tfrac{1}{3},$$

since outward normal is in the $-\mathbf{j}$ direction. For the plane ABC, the normal is in the direction of the vector with components $\mathbf{n} = (2, 2, 1)$. Hence $\mathbf{n} \cdot \mathbf{k} = 1$. Now

$$\mathbf{n} \cdot \mathbf{F} = 2(x-2) + 2(x+3y) + 2z = 2(2x + 3y + z - 2) = 2y,$$

using the equation of the plane. Hence for the plane

$$\iint_S \mathbf{F} \cdot d\mathbf{S} = 2 \iint y \, dx \, dy = 2 \int_{y=0}^{1} y \, dy \left[\int_0^{1-y} dx \right]$$

$$= 2 \int_0^1 y(1-y) \, dy = \tfrac{1}{3}.$$

The sum of these four results gives the outward flux $\iint_S \mathbf{F} \cdot d\mathbf{S}$ as
$0 + 2 - \tfrac{1}{3} + \tfrac{1}{3} = 2.$

By the divergence theorem div $\mathbf{F} = 6$, so

$$\iiint_V \operatorname{div} \mathbf{F} \, dV = 6 \times \text{volume of tetrahedron } (= \tfrac{1}{3}).$$

Hence again

$$\iint_S \mathbf{F} \cdot d\mathbf{S} = 6 \times \tfrac{1}{3} = 2.$$

21. PARTIAL DIFFERENTIAL EQUATIONS

[Formation of partial differential equations, classification: elliptic, parabolic, hyperbolic. Separation of variables, spherical and cylindrical coordinate systems, Laplace transform method, similarity solutions, non-homogeneous equations and boundary conditions, characteristics]

21.1 Eliminate the arbitrary functions from the following, so obtaining partial differential equations of which the general solutions are:

(a) $u = f(x+y)$, (b) $u = f\left(\dfrac{x}{y}\right)$, (c) $u = f(x+y) + g(x-y)$,

(d) $u = yf(x) + [f(x)]^2$.

In the following f' denotes the differential of the function f with respect to its argument.

(a) $\dfrac{\partial u}{\partial x} = \left[\dfrac{\partial}{\partial x}(x+y)\right]f' = f'$, $\dfrac{\partial u}{\partial y} = \left[\dfrac{\partial}{\partial y}(x+y)\right]f' = f'$.

Hence $\partial u/\partial x = \partial u/\partial y$.

(b) $\dfrac{\partial u}{\partial x} = \dfrac{1}{y}f'$, $\dfrac{\partial u}{\partial y} = -\dfrac{x}{y^2}f'$. Hence $x\dfrac{\partial u}{\partial x} + y\dfrac{\partial u}{\partial y} = 0$.

(c) $\dfrac{\partial u}{\partial x} = f' + g'$, $\dfrac{\partial^2 u}{\partial x^2} = f'' + g''$, $\dfrac{\partial u}{\partial y} = f' - g'$, $\dfrac{\partial^2 u}{\partial y^2} = f'' + g''$.

Hence $\partial^2 u/\partial x^2 = \partial^2 u/\partial y^2$.

(d) $\dfrac{\partial u}{\partial y} = f.$ Hence $u = y\dfrac{\partial u}{\partial y} + \left(\dfrac{\partial u}{\partial y}\right)^2$.

21.2 Find the general solution of

$$\dfrac{\partial^2 \phi}{\partial x\, \partial y} + \dfrac{\partial \phi}{\partial y} = 0.$$

Integrating with respect to y gives $(\partial\phi/\partial x)+\phi = f(x)$. The integrating factor for this equation is e^x (see 16.4(d)). Hence

$$\frac{\partial}{\partial x}(e^x\phi) = e^x\frac{\partial\phi}{\partial x}+e^x\phi = e^x f(x)$$

and therefore

$$e^x\phi = \int e^x f(x)\,dx + g(y) \quad \text{or} \quad \phi = e^{-x}(g(x)+h(y)),$$

where f, g and h are arbitrary functions.

21.3 Find the solution of $\partial^2 u/\partial x\partial y = 0$ such that $u = 1 + y$ on the y-axis and $u = 1 + x^2$ on the x-axis.

Integrating first with respect to x and then with respect to y, we have $u = F(x)+G(y)$, where $F(x)$ and $G(y)$ are arbitrary functions. For $x = 0$, $u = 1+y$. Hence $1+y = F(0)+G(y)$. For $y = 0$, $u = 1+x^2$ so $1+x^2 = G(0)+F(x)$. Hence

$$G(y) = y+1-F(0),$$
$$F(x) = x^2+1-G(0),$$

and therefore $F(0)+G(0) = 1$ (from either of these last two equations). The solution is therefore

$$u = F(x)+G(y) = x^2+y+2-(F(0)+G(0)) = x^2+y+1.$$

21.4 Solve the equation

$$\frac{\partial^2 u}{\partial x^2}+3\frac{\partial^2 u}{\partial x\,\partial y}+2\frac{\partial^2 u}{\partial y^2} = 12y.$$

The solution of the corresponding homogeneous equation (right-hand side $= 0$) is obtained by transforming to the canonical form

$$\frac{\partial^2 u}{\partial\xi\,\partial\eta} = 0,$$

by writing $\xi = x+\lambda_1 y$, $\eta = x+\lambda_2 y$, where λ_2, λ_2 are the roots of the equation $1+3\lambda+2\lambda^2 = 0$. Hence $\lambda_1 = -1$, $\lambda_2 = -\frac{1}{2}$. Therefore

$$u = f(x-y)+g(x-\tfrac{1}{2}y) \quad (f \text{ and } g \text{ are arbitrary functions}).$$

A particular integral of the inhomogeneous equation is (by observation) $u = y^3$. Hence the general solution is

$$u = f(x-y)+g(x-\tfrac{1}{2}y)+y^3.$$

21.5 State the nature of the following equations, and obtain their general solutions:

(a) $3\dfrac{\partial^2 u}{\partial x^2}+4\dfrac{\partial^2 u}{\partial x\,\partial y}-\dfrac{\partial^2 u}{\partial y^2}=0,$

(b) $\dfrac{\partial^2 u}{\partial x^2}-2\dfrac{\partial^2 u}{\partial x\,\partial y}+\dfrac{\partial^2 u}{\partial y^2}=0,$

(c) $4\dfrac{\partial^2 u}{\partial x^2}+\dfrac{\partial^2 u}{\partial y^2}=0.$

The Euler equation

$$a\frac{\partial^2 u}{\partial x^2}+2h\frac{\partial^2 u}{\partial x\,\partial y}+b\frac{\partial^2 u}{\partial y^2}=0$$

is classified as being:

$\begin{cases} \text{hyperbolic if } ab-h^2<0, \text{ general solution } u=f(x+\lambda_1 y)+g(x+\lambda_2 y)\\ \text{parabolic if}\quad ab-h^2=0, \text{ general solution } u=f(x+\lambda_1 y)+xg(x+\lambda_1 y)\\ \text{elliptic if}\qquad ab-h^2>0, \text{ general solution } u=f(x+\lambda_1 y)+g(x+\lambda_2 y). \end{cases}$

where λ_1, λ_2 are the roots of the equation $a+2h\lambda+b\lambda^2=0$. (N.B. in the parabolic case this quadratic has two identical roots, λ_1, say.)

(a) $a=3$, $b=-1$, $h=2$, $ab-h^2=-7$. Hence hyperbolic. Also $3+4\lambda-\lambda^2=0$ gives $\lambda=2\pm\sqrt{7}$. The general solution is therefore

$u=f(x+(2+\sqrt{7})y)+g(x+(2-\sqrt{7})y).$

(b) $a=1$, $b=1$, $h=-1$, $ab^2-h^2=0$. Hence parabolic. Also $1-2\lambda+\lambda^2=0$ gives $\lambda=1$ (twice). The general solution is therefore

$u=f(x+y)+xg(x+y).$

(c) $a=4$, $b=1$, $h=0$, $ab-h^2=4$ (>0). Hence elliptic. Now $4+\lambda^2=0$ gives $\lambda=\pm2i$. The general solution is therefore

$u=f(x+2iy)+g(x-2iy).$

21.6 Find separable solutions of $\dfrac{\partial^2 u}{\partial x^2}=\dfrac{1}{k}\dfrac{\partial u}{\partial t}$ (k constant) satisfying the following sets of boundary conditions:

(a) $u=0$ when $x=0$ and $x=\pi$, and $u\to 0$ as $t\to\infty$.

(b) $\dfrac{\partial u}{\partial x}=0$ when $x=0$ and $x=\pi$, and $u\to 0$ as $t\to\infty$.

(a) Let $u=X(x)T(t)$. Then $\dfrac{X''}{X}=\dfrac{1}{k}\dfrac{T'}{T}=-\lambda^2$ (the negative sign being chosen so that $T \to 0$ as $t \to \infty$). Then

$$T = Ce^{-\lambda^2 kt}, \qquad X = A \cos \lambda x + B \sin \lambda x.$$

The boundary conditions give $A = 0$, $\sin \lambda \pi = 0$ or $\lambda = n$ $(n = 1, 2, 3, \ldots)$. Hence $u(x, t) = Ce^{-n^2 kt} \sin nx$ (B being absorbed into the constant C).

(b) The same solutions for X and T apply, but the boundary conditions now give (since $dX/dx = 0$ at $x = 0$ and $x = \pi$) $B = 0$, $\sin \lambda \pi = 0$ or $\lambda = n$ $(n = 1, 2, 3, \ldots)$. Hence

$$u(x, t) = Ce^{-n^2 kt} \cos nx.$$

21.7 Given that x and y are independent variables and $u(t)$ is a function of t only, where $t = x^2 + y^2$, show that $\nabla^2 u = 4t\, d^2 u/dt^2 + 4\, du/dt$. Hence show that $u = A \log_e (x^2 + y^2) + B$ is a solution of Laplace's equation $\nabla^2 u = 0$, where A and B are arbitrary constants.

$$\frac{\partial u}{\partial x} = \frac{du}{dt}\frac{\partial t}{\partial x}, \qquad \frac{\partial^2 u}{\partial x^2} = 4x^2 \frac{d^2 u}{dt^2} + 2\frac{du}{dt}, \qquad \frac{\partial^2 u}{\partial y^2} = 4y^2 \frac{d^2 u}{dt^2} + 2\frac{du}{dt}.$$

Therefore

$$\nabla^2 u = 4t \frac{d^2 u}{dt^2} + 4\frac{du}{dt}.$$

Hence $\nabla^2 u = 0$ gives $u''/u' = -1/t$. Integrating twice gives

$$u = A \log_e t + B = A \log_e (x^2 + y^2) + B.$$

21.8 A function Φ satisfies the equation

$$\Phi_{xx} + \Phi_{yy} = a^2 \Phi,$$

where a is a real constant, in the interior of the unit square OABC with vertices $(0, 0)$, $(0, 1)$, $(1, 0)$, $(1, 1)$. Φ vanishes on the sides OA, OB, and BC, while on the side AC $(y = 1)$ Φ equals $\sin n\pi x$, where n is an integer. Show that at the centre of the square

$$\Phi(\tfrac{1}{2}, \tfrac{1}{2}) = \left(\sin \frac{n\pi}{2} \right) \frac{\sinh \tfrac{1}{2} p_n}{\sinh p_n},$$

where $p_n^2 = a^2 + n^2 \pi^2$. (L.U.)

Put $\Phi = X(x)T(t)$. Then $X''/X = -\lambda^2$, $Y''/Y = a^2 + \lambda^2$, where λ is a constant (the negative sign in the X equation being chosen to give periodic solutions in x as required by the boundary condition on $y = 1$). Hence solving these two ordinary differential equations we have

$$X = A \cos \lambda x + B \sin \lambda x,$$
$$Y = C \cosh \sqrt{a^2 + \lambda^2}\, y + D \sinh \sqrt{a^2 + \lambda^2}\, y$$

(A, B, C, D are integration constants).
Now along OA: $x = 0$, $\Phi = 0$. Therefore $A = 0$.
Along OB: $y = 0$, $\Phi = 0$, and hence $C = 0$.
Along BC: $x = 1$, $\Phi = 0$. Therefore $\sin \lambda = 0$, which implies $\lambda = n\pi$, $n = 1, 2, 3, \ldots$.
Using the values for the constants obtained from these boundary conditions

$$\Phi = BD \sin n\pi x \sinh \sqrt{a^2 + n^2\pi^2}\, y$$

or $\Phi = E \sin n\pi x \sinh \sqrt{a^2 + n^2\pi^2}\, y$ (where E is an integration constant). Using the condition along AC ($\Phi = \sin n\pi x$ along $y = 1$) we find

$$E = \frac{1}{\sinh \sqrt{a^2 + n^2\pi^2}} = \frac{1}{\sinh p_n},$$

where

$$p_n^2 = a^2 + n^2\pi^2.$$

Hence

$$\Phi = \frac{\sin n\pi x \sinh p_n y}{\sinh p_n}.$$

At $(\tfrac{1}{2}, \tfrac{1}{2})$ then

$$\Phi(\tfrac{1}{2}, \tfrac{1}{2}) = \left(\sin \frac{n\pi}{2} \right) \frac{\sinh (p_n/2)}{\sinh p_n}.$$

21.9 Transform the equation

$$x^2 \frac{\partial^2 u}{\partial x^2} = \frac{1}{c^2} \frac{\partial^2 u}{\partial t^2},$$

where c is a constant, into a constant coefficient equation by the transformation $x = e^z$, and hence by separation of variables find the solution $u(x, t)$, periodic in t, which satisfies the conditions $u(x, t) = 0$ at $x = l$ and $x = 2l$, where l is a constant.

$$\frac{\partial u}{\partial x} = \frac{\partial u}{\partial z}\frac{dz}{dx} = \frac{1}{x}\frac{\partial u}{\partial z}; \qquad \frac{\partial^2 u}{\partial x^2} = \frac{\partial}{\partial x}\left(\frac{1}{x}\frac{\partial u}{\partial z} \right) = \frac{1}{x^2}\frac{\partial^2 u}{\partial z^2} - \frac{1}{x^2}\frac{\partial u}{\partial z}.$$

Hence the equation becomes

$$\frac{\partial^2 u}{\partial z^2} - \frac{\partial u}{\partial z} = \frac{1}{c^2} \frac{\partial^2 u}{\partial t^2}.$$

Now putting $u(z, t) = Z(z)T(t)$ we have

$$\frac{Z''}{Z} - \frac{Z'}{Z} = \frac{1}{c^2} \frac{T'}{T} = -\frac{\omega^2}{c^2}.$$

where ω is a real constant, the negative sign of the separation constant being chosen so that solutions in t are periodic as required.

The constant coefficient equation for Z has as its solution (obtained by writing $Z = e^{mz}$ and solving for m in the usual way – see 16.7)

$$Z = e^{z/2} \left[E \cos \left(\frac{1}{2} \sqrt{\frac{4\omega^2}{c^2} - 1} \right) z + F \sin \left(\frac{1}{2} \sqrt{\frac{4\omega^2}{c^2} - 1} \; z \right) \right],$$

where E and F are arbitrary constants of integration. Hence

$$u(x, t) = x^{1/2} \left[E \cos \left(\frac{1}{2} \sqrt{\frac{4\omega^2}{c^2} - 1} \; \log_e x \right) \right.$$

$$\left. + F \sin \left(\frac{1}{2} \sqrt{\frac{4\omega^2}{c^2} - 1} \; \log_e x \right) \right] (A \cos \omega t + B \sin \omega t).$$

Now $u = 0$ at $x = l$. Hence

$$0 = E \cos \left(\frac{1}{2} \sqrt{\frac{4\omega^2}{c^2} - 1} \; \log_e l \right) + F \sin \left(\frac{1}{2} \sqrt{\frac{4\omega^2}{c^2} - 1} \; \log_e l \right).$$

Similarly $u = 0$ at $x = 2l$ gives

$$0 = E \cos \left(\frac{1}{2} \sqrt{\frac{4\omega^2}{c^2} - 1} \; \log_e 2l \right) + F \sin \left(\frac{1}{2} \sqrt{\frac{4\omega^2}{c^2} - 1} \; \log_e 2l \right).$$

These last two equations have the trivial solution $E = F = 0$, but *non-trivial* solutions exist provided the determinant of the coefficients of E and F vanishes, that is

$$\begin{vmatrix} \cos (K \log_e l) & \sin (K \log_e l) \\ \cos (K \log_e 2l) & \sin (K \log_e 2l) \end{vmatrix} = 0,$$

where $K = \frac{1}{2}\sqrt{(4\omega^2/c^2) - 1}$. This gives, after some simplification.

$$\sin (K \log_e 2) = 0.$$

177

Hence

$$K = \frac{1}{2}\sqrt{\frac{4\omega^2}{c^2} - 1} = \frac{n\pi}{\log_e 2} \qquad (n = 1, 2, 3, \ldots),$$

and therefore

$$\omega^2 = c^2 \left[\frac{n^2\pi^2}{(\log_e 2)^2} + \frac{1}{4} \right].$$

Accordingly

$$\frac{E}{F} = -\tan\left(\frac{n\pi \log_e l}{\log_e 2}\right).$$

Inserting this ratio into the solution for $u(x, t)$ gives

$$u(x, t) = x^{1/2} \sin\left(\frac{n\pi}{\log_e 2} \log_e \frac{x}{l}\right)(A \cos \omega t + B \sin \omega t),$$

where A and B are arbitrary constants.

21.10 Solve the equation

$$\frac{\partial^2 u}{\partial x^2} = \frac{1}{k}\frac{\partial u}{\partial t}$$

subject to the conditions $u = 0$ at $x = 0$, $\partial u/\partial x = 0$ at $x = a$, and $u = u_0 \sin^3(\pi x/2a)$ at $t = 0$, where k (>0), u_0 and a are constants.

Put $u(x, t) = X(x)T(t)$.
Then

$$\frac{X''}{X} = \frac{1}{k}\frac{T'}{T} = -\omega^2,$$

whence $X = A \cos \omega x + B \sin \omega x$, $T = e^{-k\omega^2 t}$, and therefore

$$u(x, t) = (A \cos \omega x + B \sin \omega x)e^{-k\omega^2 t}.$$

Now $u = 0$ at $x = 0$. Therefore $A = 0$ and hence

$$u(x, t) = Be^{-k\omega^2 t} \sin \omega x.$$

Also $\dfrac{\partial u}{\partial x} \equiv \dfrac{dX}{dx} = 0$ at $x = a$. Therefore $\cos \omega a = 0$, or $\omega = \left(\dfrac{2n+1}{2}\right)\dfrac{\pi}{a}$ where $n = 0, 1, 2, 3, 4, \ldots$. Hence

$$u(x, t) = \sum_{n=0}^{\infty} B_n \exp\left[-k\left(\frac{2n+1}{2}\right)^2 \frac{\pi^2}{a^2} t\right]$$

$$\times \sin\left(\frac{2n+1}{2}\right)\frac{\pi x}{a}.$$

Now

$$\sin^3 \theta = \tfrac{1}{2}\sin \theta(1 - \cos 2\theta) = \tfrac{1}{2}\sin \theta - \tfrac{1}{4}(\sin 3\theta - \sin \theta)$$
$$= \tfrac{3}{4}\sin \theta - \tfrac{1}{4}\sin 3\theta.$$

Hence at $t = 0$

$$\frac{3}{4}u_0 \sin \frac{\pi x}{2a} - \frac{1}{4}u_0 \sin \frac{3\pi x}{2a} = \sum_{n=0}^{\infty} B_n \sin\left(\frac{2n+1}{2}\right)\frac{\pi x}{a}.$$

Comparing coefficients we find

$$B_0 = \tfrac{3}{4}u_0, \qquad B_1 = -\tfrac{1}{4}u_0, \qquad B_n = 0 \qquad (n > 1),$$

and therefore

$$u(x, t) = \frac{u_0}{4}\left\{3\mathrm{e}^{-\pi^2 kt/(4a^2)} \sin \frac{\pi x}{2a} - \mathrm{e}^{-9\pi^2 kt/(4a^2)} \sin \frac{3\pi x}{2a}\right\}.$$

21.11 The function $u(x, y)$ satisfies the equation

$$\frac{\partial^2 u}{\partial x^2} + \frac{\partial^2 u}{\partial y^2} = 0$$

in the region $0 \le x \le l$, $0 \le y < \infty$, and is zero on the boundary except for $y = 0$, $0 < x < l$, where it takes a constant value u_0. Show that

$$u(x, y) = \frac{4u_0}{\pi} \sum_{\text{odd } n} \frac{1}{n}\mathrm{e}^{-n\pi y/l} \sin \frac{n\pi x}{l},$$

and deduce that at $x = \tfrac{1}{2}l$

$$u\left(\frac{l}{2}, y\right) = \frac{4u_0}{\pi} \tan^{-1}(\mathrm{e}^{-\pi y/l}).$$

Put $u(x, y) = X(x)Y(y)$. In the usual way this gives

$$u(x, y) = \sum_{n=1}^{\infty} A_n \mathrm{e}^{-\pi n y/l} \sin \frac{n\pi x}{l},$$

where the A_n are a set of constants yet to be found. Now on $y = 0$, $0 < x < l$,

$$u_0 = \sum_{n=1}^{\infty} A_n \sin \frac{n\pi x}{l}.$$

This is a Fourier sine series and the coefficients A_n are just the appropriate Fourier coefficients (see 19.2) given by

$$A_n = \frac{2}{l} \int_0^l u_0 \sin \frac{n\pi x}{l} \, \mathrm{d}x$$
$$= \begin{cases} 4u_0/n\pi, & n = 1, 3, 5, \ldots \\ 0, & n = 2, 4, 6, \ldots \end{cases}$$

Hence

$$u(x, y) = \frac{4u_0}{\pi} \sum_{\text{odd } n}^{\infty} \frac{1}{n} e^{-n\pi y/l} \sin \frac{n\pi x}{l}.$$

At

$$x = \tfrac{1}{2}l, \qquad u\left(\frac{l}{2}, y\right) = \frac{4u_0}{\pi} (e^{-\pi y/l} - \tfrac{1}{3}e^{-3\pi y/l} + \tfrac{1}{5}e^{-5\pi y/l} - \cdots).$$

Now the Maclaurin expansion of $\tan^{-1} x$ gives

$$\tan^{-1} x = x - \frac{x^3}{3} + \frac{x^5}{5} - \cdots \quad \text{for} \quad |x| < 1.$$

Since $|e^{-\pi y/l}| < 1$, we have finally

$$u\left(\frac{l}{2}, y\right) = \frac{4u_0}{\pi} \tan^{-1} (e^{-\pi y/l}).$$

21.12 Find separable solutions of the equation

$$\frac{\partial^4 u}{\partial x^4} + \frac{1}{k^2} \frac{\partial^2 u}{\partial t^2} = 0$$

in the form $u = X(x)T(t)$, and show that $T(t)$ can be written as $C \sin(\omega^2 kt + \alpha)$, where C, ω and α are constants. Show that if solutions of this type satisfy the boundary conditions $u = \partial u/\partial x = 0$ at $x = 0$ and $x = l$, then

$$\cos \omega l \cosh \omega l = 1.$$

Put $u = X(x)T(t)$. Then

$$\frac{X'''}{X} = -\frac{1}{k^2} \frac{T''}{T} = \omega^4,$$

where ω is a separation constant. Hence $T'' + k^2\omega^4 T = 0$, which has as its solution $T = C \sin(\omega^2 kt + \alpha)$, where C and α are arbitrary constants. Now

$$X'''' - \omega^4 X = 0$$

gives

$$X = A_1 \cos \omega x + A_2 \sin \omega x + B_1 \cosh \omega x + B_2 \sinh \omega x,$$

where A_1, A_2, B_1, B_2 are arbitrary constants. The boundary condition $u = 0$ at $x = 0$ gives $A_1 + B_1 = 0$. Likewise $\partial u/\partial x = 0$ at $x = 0$ gives $A_2 + B_2 = 0$. Similarly $u = 0$ at $x = l$ gives

$$A_1 \cos \omega l + A_2 \sin \omega l + B_1 \cosh \omega l + B_2 \sinh \omega l = 0,$$

and $\partial u/\partial x = 0$ at $x = l$ gives

$$-A_1 \sin \omega l + A_2 \cos \omega l + B_1 \sinh \omega l + B_2 \cosh \omega l = 0.$$

Inserting in these last two equations $B_1 = -A_1$ and $B_2 = -A_2$, we have the pair of equations

$$A_1(\cos \omega l - \cosh \omega l) + A_2(\sin \omega l - \sinh \omega l) = 0,$$

$$-A_1(\sin \omega l + \sinh \omega l) + A_2(\cos \omega l - \cosh \omega l) = 0,$$

whence for non-trivial (i.e. non-zero) solutions for the A_1 and A_2 we must have

$$\begin{vmatrix} \cos \omega l - \cosh \omega l & \sin \omega l - \sinh \omega l \\ -(\sin \omega l + \sinh \omega l) & \cos \omega l - \cosh \omega l \end{vmatrix} = 0$$

or

$$\cos \omega l \cosh \omega l = 1.$$

21.13 Show that $u = (Ar^\lambda + Br^{-\lambda})$, where A, B, λ, α are constants, is a solution of Laplace's equation in polar coordinates (r, θ)

$$\frac{\partial^2 u}{\partial r^2} + \frac{1}{r}\frac{\partial u}{\partial r} + \frac{1}{r^2}\frac{\partial^2 u}{\partial \theta^2} = 0.$$

Solve this equation in a semi-circular region bounded by the x-axis and the semi-circle $x^2 + y^2 = a^2$, $y > 0$, given that $u = 0$ on the straight boundary $-a < x < a$, and $u = u_0$ on the curved boundary, where u_0 is constant. (See Fig. 40.)

Put $u = R(r)\Theta(\theta)$. Then

$$\frac{r}{R}\frac{d}{dr}\left(r\frac{dR}{dr}\right) = -\frac{1}{\Theta}\frac{d^2\Theta}{d\theta^2} = \lambda^2 \qquad (\lambda > 0),$$

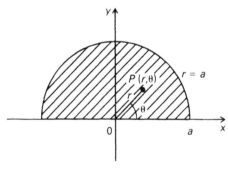

Fig. 40

where λ is a separation constant. Hence

$$\Theta = E \sin \lambda\theta + F \cos \lambda\theta = \sin(\lambda\theta + \alpha) \qquad (\alpha \text{ is a constant})$$

and

$$\frac{d}{dr}\left(r\frac{dR}{dr}\right) = \frac{\lambda^2}{r} R.$$

Writing $R = G(r)/r^{1/2}$, we have $d^2G/dr^2 = (\lambda^2 - \frac{1}{4})G/r^2$, which, by putting $G = r^p$, requires $p = \frac{1}{2} + \lambda$ or $\frac{1}{2} - \lambda$. Hence solutions of the equation for R are

$$R = r^\lambda \text{ and } r^{-\lambda}.$$

Accordingly

$$u(r, \theta) = (Ar^\lambda + Br^{-\lambda}) \sin(\lambda\theta + \alpha),$$

where A and B are arbitrary constants of integration.

To ensure a finite solution at $r = 0$ we must take $B = 0$ for all λ. The solution is therefore

$$u(r, \theta) = \sum_\lambda A_\lambda r^\lambda \sin(\lambda\theta + \alpha).$$

Now on $\theta = 0$, $0 \leq r < a$, $u = 0$. Hence $0 = \sum_\lambda A_\lambda r^\lambda \sin \alpha$, which is satisfied by choosing $\alpha = 0$. For $\theta = \pi$, $0 \leq r < a$, $u = 0$. Hence $0 = \sum_\lambda A_\lambda r^\lambda \sin \lambda\pi$, and therefore $\lambda = n \ (= 1, 2, 3, \ldots)$. The solution now takes the form

$$u(r, \theta) = \sum_{n=1}^{\infty} A_n r^n \sin n\theta.$$

Using the boundary condition on the curved part of the boundary, $u = u_0$ on $r = a$,

$$u_0 = \sum_{n=1}^{\infty} A_n a^n \sin n\theta \qquad (0 < \theta < \pi),$$

which is a Fourier sine expansion of the constant u_0, A_n being the Fourier coefficients. Therefore

$$a^n A_n = \frac{2}{\pi} \int_0^\pi u_0 \sin n\theta \, d\theta = \frac{2u_0}{n\pi}(1 - \cos n\pi)$$

$$= \begin{cases} 0, & n = 2, 4, 6, \ldots \\ 4u_0/n\pi, & n = 1, 3, 5, \ldots. \end{cases}$$

Putting $n = 2k + 1 \ (k = 0, 1, 2, \ldots)$, we have

$$a^{2k+1} A_{2k+1} = \frac{4u_0}{(2k+1)\pi}$$

and finally

$$u(r, \theta) = \sum_{k=0}^{\infty} \frac{4u_0}{(2k+1)\pi} \left(\frac{r}{a}\right)^{2k+1} \sin(2k+1)\theta$$

$$= \frac{4u_0}{\pi} \sum_{k=0}^{\infty} \frac{1}{2k+1} \left(\frac{r}{a}\right)^{2k+1} \sin(2k+1)\theta.$$

21.14 An infinite cylinder with a circular cross section has a radius 2 units. The surface of the cylinder is kept at zero temperature for all time t, and at $t = 0$ the distribution of temperature is given by $u = J_0(1.2r)$, where J_0 is the Bessel function of zero order, and where $u(r, \theta, t)$ satisfies the heat conduction equation in polar coordinates (r, θ)

$$\frac{1}{r}\frac{\partial}{\partial r}\left(r\frac{\partial u}{\partial r}\right) + \frac{1}{r^2}\frac{\partial^2 u}{\partial \theta^2} = \frac{1}{k}\frac{\partial u}{\partial t} \qquad (k \text{ constant}).$$

Show that at time t the temperature distribution is approximately given by

$$u = e^{-1.44kt}J_0(1.2r).$$

Since there is no θ variation the equation becomes

$$\frac{1}{r}\frac{\partial}{\partial r}\left(r\frac{\partial u}{\partial r}\right) = \frac{1}{k}\frac{\partial u}{\partial t}.$$

Now let $u = R(r)T(t)$, whence

$$\frac{1}{rR}\frac{d}{dr}\left(r\frac{dR}{dr}\right) = \frac{1}{kT}\frac{dT}{dt} = -\lambda^2,$$

where λ is a separation constant. Hence $T' = -\lambda^2 kT$, and therefore

$$T = \begin{cases} e^{-\lambda^2 kt} & \text{when } \lambda \neq 0, \\ C \text{ (const.)} & \text{when } \lambda = 0. \end{cases}$$

Now consider the R equation. This is

$$\frac{d^2R}{dr^2} + \frac{1}{r}\frac{dR}{dr} + \lambda^2 R = 0,$$

which is the Bessel equation (see 17.1) whose solution (finite at the origin) is

$$R = J_0(\lambda r) \qquad \text{when } \lambda \neq 0.$$

We also have

$$R = A \log_e r + B \qquad \text{when } \lambda = 0.$$

183

Hence taking a linear combination of the possible solutions we have

$$u(r, t) = (A \log_e r + B)C + \sum_{\lambda \neq 0} A_\lambda e^{-\lambda^2 kt} J_0(\lambda r),$$

where A, B and A_λ are constants.

Applying the boundary conditions, we must choose $A = 0$ for finiteness of the solution at $r = 0$. As $t \to \infty$, $u \to 0$. Therefore $C = 0$. Hence

$$u(r, t) = \sum_{\lambda \neq 0} A_\lambda e^{-\lambda^2 kt} J_0(\lambda r).$$

Now when $r = 2$, $u(2, t) = 0$, so

$$\sum_\lambda A_\lambda e^{-\lambda^2 kt} J_0(2\lambda) = 0,$$

which implies $J_0(2\lambda) = 0$.

Now the first zero of $J_0(x)$ is at $x \approx 2.4$, and therefore the corresponding $\lambda \approx 1.2$. Hence

$$u(r, 0) = J_0(1.2r) = \sum_\lambda A_\lambda J_0(\lambda r) \approx A_{1.2} J_0(1.2r),$$

taking just the first term in the series. Therefore $A_{1.2} = 1$. Hence

$$u(r, t) \approx e^{-(1.2)^2 kt} J_0(1.2r).$$

21.15 Find a finite solution of Laplace's equation $\nabla^2 u = 0$ inside a sphere of *unit* radius, given $u = 1 + \cos \theta - 3 \cos^2 \theta$ on the surface.

Laplace's equation in spherical polar coordinates (r, θ, ϕ) is

$$\frac{1}{r^2} \frac{\partial}{\partial r} \left(r^2 \frac{\partial u}{\partial r} \right) + \frac{1}{r^2 \sin \theta} \frac{\partial}{\partial \theta} \left(\sin \theta \frac{\partial u}{\partial \theta} \right) + \frac{1}{r^2 \sin^2 \theta} \frac{\partial^2 u}{\partial \phi^2} = 0.$$

Since u is not ϕ dependent, we write $u = R(r)\Theta(\theta)$. Then

$$\frac{d}{dr} \left(r^2 \frac{dR}{dr} \right) = \lambda R,$$

$$\frac{1}{\sin \theta} \frac{d}{d\theta} \left(\sin \theta \frac{d\Theta}{d\theta} \right) = -\lambda \Theta,$$

where λ is a separation constant. The Θ equation is Legendre's equation (see 17.7) and has *polynomial* solutions in $\cos \theta$ (as required by the boundary condition) if $\lambda = l(l+1)$, where $l = 0, 1, 2, \ldots$. The solutions are then the Legendre polynomials $P_l(\cos \theta)$, where

$$P_0 = 1, \qquad P_1 = \cos \theta, \qquad P_2 = \tfrac{1}{2}(3 \cos^2 \theta - 1).$$

The R equation now becomes

$$\frac{d}{dr}\left(r^2\frac{dR}{dr}\right) = l(l+1)R.$$

Putting $R = G/r$, we have $d^2G/dr^2 = [l(l+1)/r^2]G$, and letting $G = r^p$ we find $p = -l, l+1$. Hence

$$R(r) = \begin{cases} r^l, \\ \dfrac{1}{r^{l+1}}, \end{cases}$$

and therefore

$$u(r, \theta) = \sum_{l=0}^{\infty}\left(A_l r^l + \frac{B_l}{r^{l+1}}\right)P_l(\cos\theta),$$

where A_l and B_l are arbitrary constants of integration.

For a finite solution as $r \to 0$ we must take $B_l = 0$ for all l. Now the boundary condition only involves cosine terms up to $\cos^2\theta$. Hence we can put $A_l = 0$ for $l \geq 3$. Accordingly on $r = 1$

$$1 + \cos\theta - 3\cos^2\theta = u(1, \theta) = A_0 P_0(\cos\theta)$$
$$+ A_1 P_1(\cos\theta) + A_2 P_2(\cos\theta)$$
$$= A_0 + A_1\cos\theta + \frac{A_2}{2}(3\cos^2\theta - 1).$$

Therefore comparing corresponding terms on each side we have

$$A_0 = 0, \qquad A_1 = 1, \qquad A_2 = -2,$$

and the solution is

$$u(r, \theta) = rP_1(\cos\theta) - 2r^2 P_2(\cos\theta).$$

21.16 Use the Laplace transform method to solve the equation

$$\frac{\partial u}{\partial t} + \frac{\partial}{\partial x}[(1-x^2)u] = \frac{1}{1-x^2}$$

in $-1 \leq x \leq 1$ with the initial conditions $u(x, 0) = 0$ and $u(0, t) = t$.

Transforming with respect to t gives (using results similar to 18.4)

$$p\bar{u}(x, p) - u(x, 0) + \frac{d}{dx}[(1-x^2)\bar{u}(x, p)] = \frac{1}{p(1-x^2)}.$$

Now $u(x, 0) = 0$. Therefore

$$p\bar{u} + \frac{d}{dx}[(1-x^2)\bar{u}] = \frac{1}{p(1-x^2)}$$

or

$$\frac{d\bar{u}}{dx} + \left(\frac{p-2x}{1-x^2}\right)\bar{u} = \frac{1}{p(1-x^2)^2}.$$

The integrating factor for this equation is

$$R = \exp\left[\int\left(\frac{p-2x}{1-x^2}\right)dx\right] \qquad \text{(see 16.4(d))}$$

and the solution is given by

$$\bar{u}R = \int\frac{R\,dx}{p(1-x^2)}.$$

Now by partial fractions

$$\int\left(\frac{p-2x}{1-x^2}\right)dx = \int\left(\frac{(p/2)+1}{1+x} + \frac{(p/2)-1}{1-x}\right)dx = \log_e\left\{\frac{(1+x)^{(p/2)+1}}{(1-x)^{(p/2)-1}}\right\}.$$

Hence

$$R = (1-x^2)\left(\frac{1+x}{1-x}\right)^{p/2},$$

and therefore

$$\bar{u}\left(\frac{1+x}{1-x}\right)^{p/2}(1-x^2) = \frac{1}{p}\int\left(\frac{1+x}{1-x}\right)^{p/2}\frac{1}{1-x^2}\,dx$$

$$= \frac{1}{p^2}\left(\frac{1+x}{1-x}\right)^{p/2} + C.$$

Using $u(0, t) = t$ we have $\bar{u}(0, p) = 1/p^2$. Hence $C = 0$, and therefore $\bar{u}(x, p) = 1/[p^2(1-x^2)]$.

Inverting gives the solution

$$u(x, t) = \frac{t}{1-x^2}.$$

21.17 By taking the Laplace transform with respect to t, show that the equation

$$\frac{\partial^2\phi}{\partial x^2} = \frac{\partial\phi}{\partial t} \qquad \text{for } 0 \leqslant x < a,$$

where $a \neq n\pi$, $n = 1, 2, 3, \ldots$, subject to the boundary conditions

$$\phi(0, t) = 0, \quad \phi(a, t) = 0 \qquad \text{for } t \geqslant 0,$$
$$\phi(x, 0) = \sin x \qquad \text{for } 0 \leqslant x < a,$$

leads to

$$\bar{\phi}(x, p) = -\left(\frac{\sinh \sqrt{p}\,x}{\sinh \sqrt{p}\,a}\right)\left(\frac{\sin a}{1+p}\right) + \frac{\sin x}{1+p}.$$

Given that

$$L^{-1}\left\{\frac{\sinh \sqrt{p}\,x}{\sinh \sqrt{p}\,a}\right\} = \frac{2\pi}{a^2} \sum_{n=1}^{\infty} (-1)^n n e^{-n^2\pi^2 t/a^2} \sin \frac{n\pi x}{a}$$

use the convolution theorem

$$L^{-1}\{\bar{f}(p)\bar{g}(p)\} = \int_0^t f(t-u)g(u)\,du$$

to obtain ϕ as an infinite series.　　　　　　　　　　　　(L.U.)

Take the Laplace transform with respect to t so that

$$\frac{d^2\bar{\phi}}{dx^2} = p\bar{\phi} - \phi(x, 0) = p\bar{\phi} - \sin x.$$

Hence

$$\bar{\phi} = A \cosh \sqrt{p}\,x + B \sinh \sqrt{p}\,x + \frac{\sin x}{1+p},$$

(using the D-operator method of 16.7 to obtain the particular integral), A and B being integration constants. Now

$$\phi(0, t) = 0, \qquad \bar{\phi}(0, p) = 0,$$
$$\phi(a, t) = 0, \qquad \bar{\phi}(a, p) = 0.$$

Hence $A = 0$ and

$$B \sinh \sqrt{p}\,a = -\frac{\sin a}{1+p}.$$

Therefore

$$\bar{\phi}(x, p) = -\left(\frac{\sin a}{1+p}\right)\left(\frac{\sinh \sqrt{p}\,x}{\sinh \sqrt{p}\,a}\right) + \frac{\sin x}{1+p}.$$

Inverting we find

$$\phi(x, t) = -\sin a L^{-1}\left\{\left(\frac{\sinh \sqrt{p}\,x}{\sinh \sqrt{p}\,a}\right)\left(\frac{1}{1+p}\right)\right\} + e^{-t}\sin x.$$

which, using the convolution theorem and the given inversion of

$\sinh \sqrt{p}\, x / \sinh \sqrt{p}\, a$, gives

$$\phi(x, t) = -\sin a \int_0^t e^{-u} \frac{2\pi}{a^2} \sum_{n=1}^{\infty} n(-1)^n \exp\left[-\frac{n^2\pi^2}{a^2}(t-u)\right]$$

$$\times \sin \frac{n\pi x}{a}\, du + e^{-t} \sin x$$

$$= -\frac{2\pi \sin a}{a^2} \sum_{n=1}^{\infty} n(-1)^n \sin \frac{n\pi x}{a} \left\{ \int_0^t \exp\left[\left(\frac{n^2\pi^2}{a^2}-1\right)u\right] du \right\}$$

$$\times \exp\left[\left(-\frac{n^2\pi^2 t}{a^2}\right)\right] + \exp(-t)\sin x$$

$$= -\frac{2\pi \sin a}{a^2} \sum_{n=1}^{\infty} \left[n(-1)^n \sin \frac{n\pi x}{a} \exp\left(-\frac{n^2\pi^2}{a^2}t\right) \right.$$

$$\left. \times \left\{ \frac{\exp\left[\left(\frac{n^2\pi^2}{a^2}-1\right)t\right]-1}{(n^2\pi^2/a^2)-1} \right\} \right] + \exp(-t)\sin x, \qquad (a \neq n\pi).$$

21.18 Find the bounded solution of the wave equation

$$\frac{\partial^2 u}{\partial x^2} = \frac{1}{c^2}\frac{\partial^2 u}{\partial t^2}$$

which satisfies the boundary conditions $u(x, 0) = 0$, $(\partial u/\partial t)_{t=0} = 0$, and $u(0, t) = J_0(at)$, where a and c are constants, and J_0 is the zero-order Bessel function.

Taking the Laplace transform with respect to t we have

$$\frac{d^2\bar{u}}{dx^2} = \frac{1}{c^2}\left(p^2\bar{u}(x, p) - pu(x, 0) - \left(\frac{\partial u}{\partial t}\right)_{t=0}\right) = \frac{p^2}{c^2}\bar{u}(x, p).$$

Hence

$$\bar{u}(x, p) = Ae^{-(p/c)x} + Be^{(p/c)x},$$

where A and B are integration constants.

For a bounded solution for large x, B must be chosen to be zero. Hence

$$\bar{u}(x, p) = Ae^{-(p/c)x}.$$

When $x = 0$, $u(0, t) = J_0(at)$. Therefore

$$A = \bar{u}(0, p) = \int_0^{\infty} e^{-pt} J_0(at)\, dt$$

and hence

$$\bar{u}(x, p) = \int_0^\infty e^{-p(t+x/c)} J_0(at)\, dt.$$

Writing $t + (x/c) = s$,

$$\bar{u}(x, p) = \int_{x/c}^\infty e^{-ps} J_0\left[a\left(s - \frac{x}{c}\right)\right] ds$$

$$= \int_0^\infty e^{-ps} H\left(s - \frac{x}{c}\right) J_0\left[a\left(s - \frac{x}{c}\right)\right] ds$$

$$= L\left\{H\left(t - \frac{x}{c}\right) J_0\left[a\left(t - \frac{x}{c}\right)\right]\right\} \text{ (dummy } s \text{ replaced by } t\text{).}$$

Hence inverting

$$u(x, t) = L^{-1}L\left\{H\left(t - \frac{x}{c}\right) J_0\left[a\left(t - \frac{x}{c}\right)\right]\right\} = H\left(t - \frac{x}{c}\right) J_0\left[a\left(t - \frac{x}{c}\right)\right].$$

21.19 Show that the diffusion equation $\partial T/\partial t = \partial^2 T/\partial x^2$ has a solution of the form $T(x, t) = t^m f(x/2\sqrt{t})$, provided $f(z)$ satisfies the differential equation

$$\frac{d^2 f}{dz^2} + 2z \frac{df}{dz} - 4mf = 0, \qquad m \text{ constant.}$$

Hence find a power series expansion for $T(x, t)$ in terms of x and t.
(L.U.)

This is an example of a *similarity* solution.
Let $z = x/2\sqrt{t}$. Then $T = t^m f(z)$,

$$\frac{\partial T}{\partial t} = mt^{m-1} f(z) + t^m\left(-\frac{x}{4t^{3/2}}\right) f'(z),$$

$$\frac{\partial T}{\partial x} = \frac{t^m}{2\sqrt{t}} f'(z), \qquad \frac{\partial^2 T}{\partial x^2} = \frac{t^{m-1}}{4} f''(z).$$

Hence T is a solution provided

$$\frac{t^{m-1}}{4} f''(z) = mt^{m-1} f(z) - \frac{x}{4} t^{m-3/2} f'(z)$$

or

$$f'' + 2zf'(z) - 4mf = 0.$$

Let

$$f(z) = \sum_{n=0}^\infty a_n z^n.$$

Then by inserting in the equation we find the recurrence relation

$$a_{n+2} = -\frac{2(n-2m)}{(n+1)(n+2)} a_n \qquad (n = 0, 1, 2, \ldots),$$

where a_0 and a_1 are arbitrary constants. Hence

$$f(z) = a_0 \left(1 + \frac{4m}{2!} z^2 + 2^2 \cdot \frac{2m(2m-2)}{4!} z^4 + \cdots\right)$$

$$+ a_1 \left(z + \frac{2(2m-1)}{3!} z^3 + 2^2 \cdot \frac{(2m-1)(2m-3)}{5!} z^5 + \cdots\right)$$

and therefore

$$T(x, t) = t^m \left[a_0 \left(1 + \frac{4m}{2!}\left(\frac{x}{2\sqrt{t}}\right)^2 + 2^2 \cdot \frac{2m(2m-2)}{4!}\left(\frac{x}{2\sqrt{t}}\right)^4 + \cdots\right)\right.$$

$$\left. + a_1 \left(\frac{x}{2\sqrt{t}} + \frac{2(2m-1)}{3!}\left(\frac{x}{2\sqrt{t}}\right)^3 + \cdots\right)\right].$$

21.20 If the diffusion equation

$$\frac{\partial^2 u}{\partial x^2} = \frac{1}{k}\frac{\partial u}{\partial t}$$

is solved for $u(x, t)$ subject to the inhomogeneous boundary conditions

$$u(0, t) = U_0, \qquad t > 0,$$
$$u(l, t) = U_1, \qquad t > 0.$$

where U_0, U_1 and l are given constants, and

$$u(x, 0) = f(x), \qquad 0 \leq x \leq l,$$

where $f(x)$ is a given function, show that the substitution

$$u(x, t) = v(x) + w(x, t)$$

can lead to a homogeneous boundary value problem for $w(x, t)$, and to a specific form for $v(x)$.

Obtain the solution of the diffusion equation when $U_0 = 0$, $U_1 = 1$, $l = 1$ and $f(x) = 0$ in the form

$$u(x, t) = x + \frac{2}{\pi} \sum_{n=1}^{\infty} \frac{(-1)^n}{n} \sin n\pi x \, e^{-n^2\pi^2 kt}. \qquad \text{(L.U.)}$$

Now $u(x, t) = v(x) + w(x, t)$ leads to

$$\frac{d^2 v}{dx^2} + \frac{\partial^2 w}{\partial x^2} = \frac{1}{k}\frac{\partial w}{\partial t}.$$

Choosing $d^2v/dx^2 = 0$ and integrating we have $v(x) = Ax + B$. Hence

$$\frac{\partial^2 w}{\partial x^2} = \frac{1}{k}\frac{\partial w}{\partial t}$$

and

$$u(0, t) = U_0 = v(0) + w(0, t),$$
$$u(l, t) = U_1 = v(l) + w(l, t).$$

Now let $v(0) = U_0$, $v(l) = U_1$ so that $w(0, t) = w(l, t) = 0$ are homogeneous boundary conditions for $w(x, t)$. Solving for A and B using $v(0)$ and $v(l)$, we have

$$v(x) = (U_1 - U_0)\frac{x}{l} + U_0.$$

When $U_0 = 0$, $U_1 = 1$, $l = 1$, $v(x) = x$. Therefore $u(x, t) = x + w(x, t)$, where $w(x, t)$ is now the solution of the homogeneous boundary value problem:

$$\frac{\partial^2 w}{\partial x^2} = \frac{1}{k}\frac{\partial w}{\partial t},$$

$w(0, t) = w(l, t) = 0$, and $w(x, 0) = u(x, 0) - v(x) = f(x) - x = -x$ (since $f(x) = 0$).

By separation of variables it is easily found that the solution of the w-equation satisfying these boundary conditions is

$$w(x, t) = \frac{2}{\pi}\sum_{n=1}^{\infty}\frac{(-1)^n}{n}\sin n\pi x\, e^{-n^2\pi^2 kt}.$$

Hence finally

$$u(x, t) = x + \frac{2}{\pi}\sum_{n=1}^{\infty}\frac{(-1)^n}{n}\sin n\pi x\, e^{-n^2\pi^2 kt}.$$

21.21 Show that the transformation $c(r, t) = u(r, t)/r$ transforms the spherically symmetric diffusion equation for the concentration $c(r, t)$

$$\frac{1}{r^2}\frac{\partial}{\partial r}\left(r^2\frac{\partial c}{\partial r}\right) = \frac{1}{k}\frac{\partial c}{\partial t}$$

into the form

$$\frac{\partial^2 u}{\partial r^2} = \frac{1}{k}\frac{\partial u}{\partial t}.$$

where k is a constant.

A sphere of radius a is initially at a uniform concentration c_0. The concentration at its surface is suddenly changed to a constant value c_1. Find the concentration at any subsequent time within the sphere.

Put $c = u/r$. Then

$$r^2 \frac{\partial c}{\partial r} = r^2 \left(-\frac{u}{r^2} + \frac{1}{r} \frac{\partial u}{\partial r} \right).$$

Therefore

$$\frac{\partial}{\partial r} \left(r^2 \frac{\partial c}{\partial r} \right) = r \frac{\partial^2 u}{\partial r^2},$$

whence

$$\frac{\partial^2 u}{\partial r^2} = \frac{1}{k} \frac{\partial u}{\partial t}.$$

Now let $u = R(r)T(t)$ so that

$$T \frac{d^2 R}{dr^2} = \frac{R}{k} \frac{dT}{dt}.$$

Hence

$$\frac{d^2 R}{dr^2} = -\lambda^2 R, \qquad \frac{1}{kT} \frac{dT}{dt} = -\lambda^2.$$

where λ is a separation constant.

Solving these equations in the usual way we have

$$R = Ar + B, \text{ for } \lambda = 0; \quad R = \sin(\lambda r + \alpha) \quad \text{for} \quad \lambda \neq 0 \ (\alpha = \text{constant}).$$

and

$$T = \text{constant} \quad \text{for } \lambda = 0; \qquad T = e^{-k\lambda^2 t}, \quad \lambda \neq 0.$$

Hence by taking a linear combination of the possible solutions

$$c(r, t) = \frac{u(r, t)}{r} = \frac{1}{r}(Ar + B) + \frac{1}{r} \sum_{\lambda \neq 0} A_\lambda e^{-k\lambda^2 t} \sin(\lambda r + \alpha)$$

where A, B and A_λ are constants.

When $t \to \infty$, $c(r, \infty) = c_1$. Hence $A = c_1$, $B = 0$. For finite c at $r = 0$, we must take $\alpha = 0$ (N.B. $\lim_{r \to 0} (\sin \lambda r / r)$ is finite). Hence

$$c(r, t) = c_1 + \frac{1}{r} \sum_{\lambda \neq 0} A_\lambda e^{-k\lambda^2 t} \sin \lambda r.$$

Now for $r = a$, $t = 0$, $c = c_1$, and therefore

$$0 = \frac{1}{a} \sum_{\lambda \neq 0} A_\lambda \sin \lambda a,$$

whence

$$\lambda = \frac{n\pi}{a}, \quad n = 1, 2, 3, \ldots.$$

Hence

$$c(r, t) = c_1 + \frac{1}{r} \sum_{n=1}^{\infty} A_n e^{-kn^2\pi^2 t/a^2} \sin \frac{n\pi r}{a}.$$

For all $r < a$, $t = 0$, $c = c_0$ (given) so

$$c_0 = c_1 + \frac{1}{r} \sum_{n=1}^{\infty} A_n \sin \frac{n\pi r}{a},$$

and therefore (using Fourier series methods)

$$A_n = \frac{2}{a} \int_0^a (c_0 - c_1) r \sin \frac{n\pi r}{a} dr$$

$$= \frac{2(c_0 - c_1)}{n\pi} \left[\frac{a}{n\pi} \sin \frac{n\pi r}{a} - r \cos \frac{n\pi r}{a} \right]_0^a$$

$$= \frac{2a}{n\pi} (c_1 - c_0)(-1)^n.$$

Hence

$$c(r, t) = c_1 + \frac{2a(c_1 - c_0)}{r\pi} \sum_{n=1}^{\infty} \frac{(-1)^n}{n} e^{-kn^2\pi^2 t/a^2} \sin \frac{n\pi r}{a}.$$

21.22   Use the method of characteristics to solve

(a) $x^2 \dfrac{\partial u}{\partial x} + y^2 \dfrac{\partial u}{\partial y} = -2x^3$ given $u = y^2 - 2$ on $y = \dfrac{1}{x}$.

(b) $\dfrac{\partial u}{\partial x} - y \dfrac{\partial u}{\partial y} = u^2$ given $u = \dfrac{1}{x}$ on $y = xe^{-x}$ (L.U.)

(a) The characteristic method gives

$$\frac{dx}{x^2} = \frac{dy}{y^2} = \frac{du}{-2x^3}.$$

Hence from the first two terms $dy/dx = y^2/x^2$, and therefore the characteristics are

$$\frac{1}{y} - \frac{1}{x} = C,$$

where C is a constant. From $dx/x^2 = du/(-2x^3)$ we have, by integrating,

$$u = -x^2 + f(C),$$

where f is an arbitrary function of the constant C which defines the characteristic. Hence

$$u = -x^2 + f\left(\frac{1}{y} - \frac{1}{x}\right).$$

Now $u = y^2 - 2$ on $y = 1/x$. Therefore

$$y^2 - 2 = -\frac{1}{y^2} + f\left(\frac{1}{y} - y\right) \quad \text{or} \quad \left(\frac{1}{y} - y\right)^2 = f\left(\frac{1}{y} - y\right).$$

Writing $\dfrac{1}{y} - y$ as a new variable z (say), we find $f(z) = z^2$. f is therefore the function which is the square of its argument. Hence the solution is

$$u = -x^2 + \left(\frac{1}{y} - \frac{1}{x}\right)^2.$$

(b) Here

$$\frac{dx}{1} = \frac{dy}{-y} = \frac{du}{u^2},$$

whence $x = -\log_e y + \log_e C$. The characteristics are therefore $ye^x = C$. Also, by integration,

$$-\frac{1}{u} = x + f(C) = x + f(ye^x).$$

But $u = 1/x$ on $y = xe^{-x}$. Hence $f(x) = -2x$. Therefore the solution is

$$\frac{1}{u} = 2ye^x - x.$$

21.23 Determine the form of ϕ such that the equation

$$y^2 \frac{\partial^2 u}{\partial x^2} - x^2 \frac{\partial u}{\partial y^2} + x \frac{\partial u}{\partial x} - y \frac{\partial u}{\partial y} = 0$$

may be written in the form $L\phi = 0$, where L is the operator

$$y \frac{\partial}{\partial x} + x \frac{\partial}{\partial y}$$

Hence solve the equation given $u = x^2$, $\dfrac{\partial u}{\partial y} = 4$ on $y = 2$. (L.U.)

Let

$$\phi = p(x, y) \frac{\partial u}{\partial x} + q(x, y) \frac{\partial u}{\partial y}.$$

Then $L\phi = 0$ is

$$\left(y\frac{\partial}{\partial x} + x\frac{\partial}{\partial y}\right)\left(p\frac{\partial u}{\partial x} + q\frac{\partial u}{\partial y}\right) = 0.$$

Hence by choosing $p = y$, $q = -x$ and performing the partial differentiation, we obtain the given equation. ϕ therefore has the form

$$y\frac{\partial u}{\partial x} - x\frac{\partial u}{\partial y}.$$

Now $L\phi = 0$ is

$$y\frac{\partial \phi}{\partial x} + x\frac{\partial \phi}{\partial y} = 0.$$

By the method of characteristics we have

$$\frac{dx}{y} = \frac{dy}{x} = \frac{d\phi}{0},$$

so the characteristic curves are $x^2 - y^2 = C$,

and

$$\phi = f(C) = f(x^2 - y^2).$$

Now

$$\phi = y\frac{\partial u}{\partial x} - x\frac{\partial u}{\partial y}$$

and therefore, using the given boundary conditions on $y = 2$, $\phi = 2(2x) - 4x = 0$ on $y = 2$. Hence $f(x^2 - 4) = 0$, and therefore $f = 0$. It follows therefore that $\phi = 0$ everywhere.

Now

$$\phi = y\frac{\partial u}{\partial x} - x\frac{\partial u}{\partial y} = 0.$$

Therefore

$$\frac{dx}{y} = \frac{dy}{-x} = \frac{du}{0},$$

and the characteristics are $x^2 + y^2 = k$ ($k = $ constant). Hence

$$u = g(k) = g(x^2 + y^2).$$

When $y = 2$, $u = x^2$. Therefore $x^2 = g(x^2 + 4)$, and writing $x^2 + 4 = v$ (say), we have $g(v) = v - 4$. Using this form for g the solution finally becomes $u = x^2 + y^2 - 4$.

22. FUNCTIONS OF MATRICES

[Exponential, sine and cosine of a matrix, Cayley–Hamilton theorem, matrix formulation of differential equations]

22.1 Show that if A and B are constant square matrices and t is a variable then

$$e^{At}e^{Bt} = e^{(A+B)t}$$

when A and B commute.

$$e^{At} = I + At + A^2\frac{t^2}{2!} + \cdots .$$

Therefore

$$e^{At}e^{Bt} = \left(I + At + A^2\frac{t^2}{2!} + \cdots\right)\left(I + Bt + B^2\frac{t^2}{2!} + \cdots\right)$$

$$= I + (A+B)t + (A^2 + 2AB + B^2)\frac{t^2}{2!} + \cdots .$$

But

$$e^{(A+B)t} = I + (A+B)t + (A+B)^2\frac{t^2}{2!} + \cdots$$

$$= I + (A+B)t + (A^2 + BA + AB + B^2)\frac{t^2}{2!} + \cdots .$$

Each term of $e^{(A+B)t} - e^{At}e^{Bt}$ can be shown using these series to contain the factor $AB - BA$. Hence when $AB = BA$

$$e^{At}e^{Bt} = e^{(A+B)t}.$$

22.2 Show that $\sin(A+B) = \sin A \cos B + \cos A \sin B$ if A and B commute.

$$\sin A = A - \frac{A^3}{3!} + \frac{A^5}{5!} - \cdots, \qquad \cos A = I - \frac{A^2}{2!} + \frac{A^4}{4!} - \frac{A^6}{6!} + \cdots.$$

$$\sin A \cos B = \left(A - \frac{A^3}{3!} + \frac{A^5}{5!} - \cdots\right)\left(I - \frac{B^2}{2!} + \frac{B^4}{4!} - \cdots\right).$$

$$\cos A \sin B = \left(I - \frac{A^2}{2!} + \frac{A^4}{4!} - \cdots\right)\left(B - \frac{B^3}{3!} + \frac{B^5}{5!} - \cdots\right).$$

$$\sin A \cos B + \cos A \sin B = (A + B)$$
$$- \frac{1}{3!}(A^3 + 3A^2B + 3AB^2 + B^3) + \cdots.$$

$$\sin(A + B) = (A + B) - \frac{1}{3!}(A + B)^3 + \cdots$$

But

$$(A + B)^3 = (A + B)(A + B)(A + B) = (A^2 + BA + AB + B^2)(A + B)$$
$$= A^3 + BA^2 + ABA + B^2A + A^2B + BAB + AB^2 + B^3.$$

When $AB = BA$ this term becomes $A^3 + 3A^2B + 3AB^2 + B^3$, which matches the corresponding term in $\sin A \cos B + \sin B \cos A$. Similarly for the other terms. Hence the result follows.

22.3 Show that

$$e^{\left(\begin{smallmatrix} 1 & 0 \\ 0 & 2 \end{smallmatrix}\right)} = \begin{pmatrix} e & 0 \\ 0 & e^2 \end{pmatrix}.$$

$$e^A = I + A + \frac{A^2}{2!} + \frac{A^3}{3!} + \cdots, \qquad A = \begin{pmatrix} 1 & 0 \\ 0 & 2 \end{pmatrix}, \qquad A^n = \begin{pmatrix} 1 & 0 \\ 0 & 2^n \end{pmatrix},$$

where n is a positive integer.
Inserting these expressions into e^A we have

$$e^{\left(\begin{smallmatrix} 1 & 0 \\ 0 & 2 \end{smallmatrix}\right)} = \begin{pmatrix} 1 & 0 \\ 0 & 1 \end{pmatrix} + \begin{pmatrix} 1 & 0 \\ 0 & 2 \end{pmatrix} + \frac{1}{2!}\begin{pmatrix} 1 & 0 \\ 0 & 2^2 \end{pmatrix} + \cdots + \frac{1}{n!}\begin{pmatrix} 1 & 0 \\ 0 & 2^n \end{pmatrix} + \cdots$$

$$= \begin{pmatrix} 1 + 1 + \dfrac{1}{2!} + \dfrac{1}{3!} + \cdots & 0 \\ 0 & 1 + 2 + \dfrac{1}{2!}2^2 + \dfrac{1}{3!}2^3 + \cdots \end{pmatrix} = \begin{pmatrix} e & 0 \\ 0 & e^2 \end{pmatrix}.$$

22.4 Show that, if H is a Hermitian matrix, then e^{iH} is a unitary matrix.

A matrix H is Hermitian if $H^\dagger(=H^{*T}) = H$. Hence $(e^{iH})^{-1} = e^{-iH} = e^{-iH\dagger} = (e^{iH})^\dagger$, and therefore e^{iH} is unitary (since a matrix U is unitary if $U^{-1} = U^\dagger$—see 14.15.)

22.5   Given that e^A is diagonalizable by a similarity transformation, show that

$$|e^A| = e^{\mathrm{Tr}\,A}.$$

Hence show that $|e^A| = 1$ when A is a skew-symmetric matrix. N.B. $|e^A|$ denotes the determinant of the matrix e^A, and $\mathrm{Tr}\,A$ denotes the trace of A (see 14.12).

If S is a matrix such that $D = S^{-1}AS$, where D is a diagonal matrix, then $S^{-1}AS$ is a similarity transformation of A. Now

$$e^D = I + D + \frac{D^2}{2!} + \frac{D^3}{3!} + \cdots$$

$$= I + S^{-1}AS + \frac{S^{-1}A^2S}{2!} + \frac{S^{-1}A^3S}{3!} + \cdots$$

$$= S^{-1}\left[I + A + \frac{A^2}{2!} + \frac{A^3}{3!} + \cdots\right]S = S^{-1}e^A S.$$

Hence

$$|e^D| = |S^{-1}e^A S| = |SS^{-1}e^A| = |e^A|.$$

Now D has diagonal elements $\lambda_1, \lambda_2, \ldots, \lambda_n$, where $\lambda_1, \ldots, \lambda_n$ are the eigenvalues of A. But the sum of the eigenvalues of a matrix is equal to the trace of the matrix, so

$$|e^D| = e^{\lambda_1}e^{\lambda_2}\cdots e^{\lambda_n} = e^{\lambda_1 + \lambda_2 + \cdots + \lambda_n} = e^{\mathrm{Tr}\,D}.$$

Since $\mathrm{Tr}\,D = \mathrm{Tr}\,A$ under a similarity transformation, $|e^A| = e^{\mathrm{Tr}\,A}$. $\mathrm{Tr}\,A = 0$ when A is skew-symmetric. Hence the result.

22.6   Use the Cayley–Hamilton theorem to evaluate e^{At}, where

$$A = \begin{pmatrix} 0 & 1 \\ -1 & 0 \end{pmatrix}.$$

Basic result

The Cayley–Hamilton theorem states that if A is a square matrix with a characteristic polynomial $f(\lambda)$ defined by

$$f(\lambda) = |A - \lambda I|,$$

then $f(A) = 0$. (N.B. This is a matrix equation, the zero on the right-hand side of the equation being the zero *matrix* of the same order as A). For the given matrix

$$f(\lambda) = \left| \begin{pmatrix} 0 & 1 \\ -1 & 0 \end{pmatrix} - \lambda \begin{pmatrix} 1 & 0 \\ 0 & 1 \end{pmatrix} \right| = \begin{vmatrix} -\lambda & 1 \\ -1 & \lambda \end{vmatrix}.$$

Hence $f(\lambda) = \lambda^2 + 1$, and accordingly by the Cayley–Hamilton theorem

$$f(A) = A^2 + I = 0.$$

Therefore $A^2 = -I$, $A^3 = -A$, $A^4 = I$, and so on. Consequently

$$e^{At} = I + At + \frac{A^2 t^2}{2!} + \cdots$$

$$= I + At - \frac{It^2}{2!} - \frac{At^3}{3!} + \frac{It^4}{4!} + \cdots$$

$$= I\left(1 - \frac{t^2}{2!} + \frac{t^4}{4!} - \cdots \right) + A\left(t - \frac{t^3}{3!} + \frac{t^5}{5!} - \cdots \right)$$

$$= I \cos t + A \sin t.$$

22.7 Evaluate, using the Cayley–Hamilton theorem, A^{-1}, where

$$A = \begin{pmatrix} 1 & 1 & 1 \\ 0 & 2 & 1 \\ -4 & 4 & 3 \end{pmatrix}.$$

$$f(\lambda) = |A - \lambda I| = \begin{vmatrix} 1-\lambda & 1 & 1 \\ 0 & 2-\lambda & 1 \\ -4 & 4 & 3-\lambda \end{vmatrix} = -\lambda^3 + 6\lambda^2 - 11\lambda + 6.$$

Hence, by the theorem,

$$f(A) = -A^3 + 6A^2 - 11A + 6I = 0.$$

Pre-multiplying by A^{-1} we have $A^2 - 6A + 11I - 6A^{-1} = 0$, whence

$$A^{-1} = \frac{A^2 - 6A + 11I}{6}.$$

Evaluating A^2 gives

$$\begin{pmatrix} -3 & 7 & 5 \\ -4 & 8 & 5 \\ -16 & 16 & 9 \end{pmatrix},$$

and hence

$$A^{-1} = \begin{pmatrix} \frac{1}{3} & \frac{1}{6} & -\frac{1}{6} \\ -\frac{2}{3} & \frac{7}{6} & -\frac{1}{6} \\ \frac{4}{3} & -\frac{4}{3} & \frac{1}{3} \end{pmatrix}.$$

22.8 Show that the ordinary differential equation

$$\frac{d^n y}{dt^n} + a_1 \frac{d^{n-1} y}{dt^{n-1}} + a_2 \frac{d^{n-2}}{dt^{n-2}} + \cdots + a_n y = 0$$

may be expressed as the equation

$$\frac{dX}{dt} = AX,$$

where X is the column matrix

$$\begin{pmatrix} y \\ y' \\ y'' \\ \vdots \\ y^{n-1} \end{pmatrix}$$

and A is a suitable square $(n \times n)$ matrix. Find the form of A in terms of the constants $a_1, a_2, \ldots, a_n$.

We can write

$$\frac{dX}{dt} = AX$$

in the form

$$\frac{d}{dt}\begin{pmatrix} y \\ y' \\ y'' \\ \vdots \\ y^{(n-1)} \end{pmatrix} = \begin{pmatrix} 0 & 1 & 0 & \cdots & 0 \\ 0 & 0 & 1 & \cdots & 0 \\ \vdots & \vdots & \vdots & \ddots & \vdots \\ 0 & 0 & 0 & \cdots & 1 \\ -a_n & -a_{n-1} & -a_{n-2} & \cdots & -a_1 \end{pmatrix}\begin{pmatrix} y \\ y' \\ \vdots \\ y^{(n-2)} \\ y^{(n-1)} \end{pmatrix}$$

which reproduces the equation in y from the last row, other rows giving identities.

Hence A is the $(n \times n)$ matrix shown.

22.9 Show that the solution of the matrix equation

$$\frac{dX(t)}{dt} = AX(t) + F(t),$$

where $X(t)$ is an $(n \times 1)$ matrix, and A and $F(t)$ are known matrices of orders $(n \times n)$ and $(n \times 1)$ respectively, is

$$X(t) = e^{At}X(0) + \int_0^t e^{A(t-s)}F(s)\, ds.$$

Find $X(t)$ when $A = \begin{pmatrix} 0 & 1 \\ -1 & 0 \end{pmatrix}$, $F(t) = \begin{pmatrix} t \\ 1 \end{pmatrix}$ and $X(0) = \begin{pmatrix} 1 \\ 2 \end{pmatrix}$.

Writing the equation as $[I(d/dt) - A]X(t) = F(t)$, the integrating factor is e^{-At}. Hence

$$\frac{d}{dt}\left(e^{-At}X(t)\right) = e^{-At}F(t)$$

and therefore

$$e^{-At}X(t) = \int_0^t e^{-As}F(s)\,ds + C,$$

where C is a constant column matrix. Putting $t = 0$, we have $C = X(0)$. Hence

$$X(t) = e^{At}X(0) + \int_0^t e^{A(t-s)}F(s)\,ds.$$

When

$$A = \begin{pmatrix} 0 & 1 \\ -1 & 0 \end{pmatrix}, \quad e^{At} = I\cos t + A\sin t \qquad \text{(see 21.6.)}.$$

Hence

$$X(t) = (I\cos t + A\sin t)\begin{pmatrix} 1 \\ 2 \end{pmatrix} + \int_0^t [I\cos(t-s) + A\sin(t-s)]\begin{pmatrix} s \\ 1 \end{pmatrix}ds$$

$$= \begin{pmatrix} 1 & 0 \\ 0 & 1 \end{pmatrix}\begin{pmatrix} 1 \\ 2 \end{pmatrix}\cos t + \begin{pmatrix} 0 & 1 \\ -1 & 0 \end{pmatrix}\begin{pmatrix} 1 \\ 2 \end{pmatrix}\sin t$$

$$+ \int_0^t \left[\begin{pmatrix} 1 & 0 \\ 0 & 1 \end{pmatrix}\begin{pmatrix} s \\ 1 \end{pmatrix}\cos(t-s) + \begin{pmatrix} 0 & 1 \\ -1 & 0 \end{pmatrix}\begin{pmatrix} s \\ 1 \end{pmatrix}\sin(t-s)\right]ds$$

$$= \begin{pmatrix} 1 \\ 2 \end{pmatrix}\cos t + \begin{pmatrix} 2 \\ -1 \end{pmatrix}\sin t$$

$$+ \int_0^t \left[\begin{pmatrix} s \\ 1 \end{pmatrix}\cos(t-s) + \begin{pmatrix} 1 \\ -s \end{pmatrix}\sin(t-s)\right]ds.$$

Therefore

$$x_1 = \cos t + 2\sin t + \int_0^t s\cos(t-s)\,ds + \int_0^t \sin(t-s)\,ds$$

$$x_2 = 2\cos t - \sin t + \int_0^t \cos(t-s)\,ds - \int_0^t s\sin(t-s)\,ds.$$

Putting $t-s=u$ and evaluating the four integrals we have

$$\int_0^t (t-u)\cos u \, du = 1-\cos t, \qquad \int_0^t \sin u \, du = 1-\cos t,$$

$$\int_0^t \cos u \, du = \sin t, \qquad \int_0^t (t-u)\sin u \, du = t-\sin t.$$

Finally therefore

$$x_1 = \cos t + 2\sin t + 1 - \cos t + 1 - \cos t = 2 - \cos t + 2\sin t,$$
$$x_2 = 2\cos t - \sin t + \sin t - t + \sin t = 2\cos t + \sin t - t.$$

22.10 Verify that the solution of the matrix equation

$$\frac{dX(t)}{dt} = AX(t) + X(t)B,$$

where A and B are constant $(n \times n)$ matrices and $X(t)$ is a square $(n \times n)$ matrix, is

$$X(t) = e^{At}X(0)e^{Bt}.$$

Evaluate $X(t)$ when $A = \begin{pmatrix} 0 & 1 \\ -1 & 0 \end{pmatrix}$, $X(0) = \begin{pmatrix} 1 & 1 \\ 0 & 1 \end{pmatrix}$. and $B = \begin{pmatrix} 0 & -1 \\ 1 & 0 \end{pmatrix}$.

$$\frac{d}{dt}X = \frac{d}{dt}(e^{At}X(0)e^{Bt}) = Ae^{At}X(0)e^{Bt} + e^{At}X(0)Be^{Bt}.$$

Since e^{Bt} and B commute, the right-hand side is $AX + XB$ and the equation is therefore satisfied.

We now require e^{At} and e^{Bt}. Using the Cayley–Hamilton theorem to find A^n and B^n (as in 21.6), we have

$$e^{At} = I\cos t + A\sin t = \begin{pmatrix} 1 & 0 \\ 0 & 1 \end{pmatrix}\cos t + \begin{pmatrix} 0 & 1 \\ -1 & 0 \end{pmatrix}\sin t = \begin{pmatrix} \cos t & \sin t \\ -\sin t & \cos t \end{pmatrix}$$

and

$$e^{Bt} = I\cos t + B\sin t = \begin{pmatrix} \cos t & -\sin t \\ \sin t & \cos t \end{pmatrix}.$$

Hence

$$X = \begin{pmatrix} \cos t & \sin t \\ -\sin t & \cos t \end{pmatrix}\begin{pmatrix} 1 & 1 \\ 0 & 1 \end{pmatrix}\begin{pmatrix} \cos t & -\sin t \\ \sin t & \cos t \end{pmatrix}$$

$$= \begin{pmatrix} 1+\cos t \sin t & \cos^2 t \\ -\sin^2 t & 1-\cos t \sin t \end{pmatrix}.$$

22.11 Express the differential equation

$$\frac{d^2y(t)}{dt^2} + (a + b \cos \omega t)y(t) = 0,$$

where a, b and ω are constants, in the matrix form

$$\frac{dX(t)}{dt} = A(t)X(t),$$

where $X(t) = \begin{pmatrix} y(t) \\ \dot{y}(t) \end{pmatrix}$, $A(t)$ being a suitable (2×2) matrix and $\dot{y}(t) = \frac{dy(t)}{dt}$. Show that $A(t)$ may be written in the form

$$A(t) = C + Db \cos \omega t,$$

where C and D are constant matrices. By the substitution $X(t) = e^{Ct}Q(t)$, obtain the differential equation

$$\frac{dQ(t)}{dt} = b \cos \omega t (e^{-Ct}De^{Ct})Q(t).$$

Using the forms obtained for C and D, evaluate $H = e^{-Ct}De^{Ct}$, where $a = 1$. Deduce that $H\dot{Q}(t) = 0$, and hence the linear relation between the components of $\dot{Q}(t)$. (L.U.)

$$\dot{X} = \begin{pmatrix} \dot{y} \\ \ddot{y} \end{pmatrix} = \begin{pmatrix} 0 & 1 \\ -a - b\cos \omega t & \omega \end{pmatrix}\begin{pmatrix} y \\ \dot{y} \end{pmatrix} = AX \text{ reproduces the equation in}$$
y. Hence

$$A = \begin{pmatrix} 0 & 1 \\ -a - b\cos \omega t & \omega \end{pmatrix} = \begin{pmatrix} 0 & 1 \\ -a & 0 \end{pmatrix} + b \cos \omega t \begin{pmatrix} 0 & 0 \\ -1 & 0 \end{pmatrix}$$

so

$$C = \begin{pmatrix} 0 & 1 \\ -a & 0 \end{pmatrix}, \qquad D = \begin{pmatrix} 0 & 0 \\ -1 & 0 \end{pmatrix}.$$

Now

$$\frac{d}{dt}[e^{Ct}Q(t)] = Ce^{Ct}Q + e^{Ct}\dot{Q}(t) = (C + bD \cos \omega t)e^{Ct}Q(t)$$

and therefore

$$e^{Ct}\dot{Q}(t) = (bD \cos \omega t)e^{Ct}Q(t),$$

giving

$$\dot{Q}(t) = b \cos \omega t (e^{-Ct}De^{Ct})Q(t) = (b \cos \omega t)HQ(t).$$

When $a = 1$,

$$C = \begin{pmatrix} 0 & 1 \\ -1 & 0 \end{pmatrix}, \qquad C^2 = \begin{pmatrix} -1 & 0 \\ 0 & -1 \end{pmatrix}.$$

Hence

$$e^{Ct} = I \cos t + C \sin t = \begin{pmatrix} \cos t & \sin t \\ -\sin t & \cos t \end{pmatrix}$$

and

$$e^{-Ct} = \begin{pmatrix} \cos t & -\sin t \\ \sin t & \cos t \end{pmatrix}.$$

Therefore

$$H = e^{-Ct}De^{Ct} = \begin{pmatrix} \sin t \cos t & \sin^2 t \\ -\cos^2 t & -\sin t \cos t \end{pmatrix}.$$

It follows that $H^2 = 0$ and therefore $H\dot{Q}(t) = 0$. Hence.

$$\begin{pmatrix} \sin t \cos t & \sin^2 t \\ -\cos^2 t & -\sin t \cos t \end{pmatrix}\begin{pmatrix} \dot{Q}_1 \\ \dot{Q}_2 \end{pmatrix} = \begin{pmatrix} 0 \\ 0 \end{pmatrix},$$

whence

$$\dot{Q}_1 \sin t \cos t + \dot{Q}_2 \sin^2 t = 0, \qquad -\cos^2 t\, \dot{Q}_1 - \sin t \cos t\, \dot{Q}_2 = 0.$$

This gives the linear relationship between $\dot{Q}_1$ and $\dot{Q}_2$ as $\dot{Q}_1 + \tan t\, \dot{Q}_2 = 0$.

22.12 The Laplace transform of an $(n \times 1)$ column matrix $x(t)$ is denoted by

$$L\{x(t)\} = \int_0^\infty e^{-pt}x(t)\, \mathrm{d}t.$$

By transforming the differential equation $\dfrac{\mathrm{d}x(t)}{\mathrm{d}t} = Ax(t)$, where A is a constant square $(n \times n)$ matrix, show that

$$x(t) = \Phi(t)x(0),$$

where $x(0)$ is the value of $x(t)$ at $t = 0$, and $\Phi(t) = L^{-1}\{(pI - A)^{-1}\}$.
Evaluate the matrix Φ when $A = \begin{pmatrix} -1 & 1 \\ -\frac{1}{3} & \frac{1}{6} \end{pmatrix}$. 　　　　(L.U.)

Now

$$L\left\{\frac{dx(t)}{dt}\right\} = \int_0^\infty e^{-pt}\frac{dx}{dt}\,dt = p\bar{x}(p) - x(0).$$

Hence the transformed equation is

$$p\bar{x}(p) - x(0) = A\bar{x}(p)$$

or

$$(pI - A)\bar{x}(p) = x(0).$$

Therefore

$$\bar{x}(p) = (pI - A)^{-1}x(0),$$

and inverting $x(t) = L^{-1}\{(pI - A)^{-1}x(0)\} = \Phi(t)x(0)$,

where

$$\Phi(t) = L^{-1}\{pI - A)^{-1}\}.$$

Now

$$pI - A = \begin{pmatrix} p & 0 \\ 0 & p \end{pmatrix} - \begin{pmatrix} -1 & 1 \\ -\frac{1}{3} & \frac{1}{6} \end{pmatrix} = \begin{pmatrix} p+1 & -1 \\ \frac{1}{3} & p-\frac{1}{6} \end{pmatrix}$$

and

$$(pI - A)^{-1} = \frac{6}{(2p+1)(3p+1)}\begin{pmatrix} p-\frac{1}{6} & 1 \\ -\frac{1}{3} & p+1 \end{pmatrix}.$$

Inverting each term of this matrix we find

$$L^{-1}\left\{\frac{6(p-\frac{1}{6})}{(2p+1)(3p+1)}\right\} = 4e^{-t/2} - 3e^{-t/3},$$

$$L^{-1}\left\{\frac{6}{(2p+1)(3p+1)}\right\} = -6e^{-t/2} + 6e^{-t/3},$$

$$L^{-1}\left\{\frac{6(-\frac{1}{3})}{(2p+1)(3p+1)}\right\} = 2e^{-t/2} - 2e^{-t/3},$$

$$L^{-1}\left\{\frac{6(p+1)}{(2p+1)(3p+1)}\right\} = -3e^{-t/2} + 4e^{-t/3}.$$

Hence

$$\Phi(t) = \begin{pmatrix} 4e^{-t/2} - 3e^{-t/3} & -6e^{-t/2} + 6e^{-t/3} \\ 2e^{-t/2} - 2e^{-t/3} & -3e^{-t/2} + 4e^{-t/3} \end{pmatrix}.$$

23. CONTOUR INTEGRATION

[Cauchy–Riemann equations, Cauchy integral formula, Laurent series, residues, Cauchy residue theorem, contour integrals]

23.1 Given that $f(z) = u + iv$ is an analytic function, and $\mathbf{F}$ is the vector $v\mathbf{i} + u\mathbf{j}$, show that the equations div $\mathbf{F} = 0$, curl $\mathbf{F} = 0$ are equivalent to the Cauchy–Riemann equations.

$$\operatorname{div}(v\mathbf{i} + u\mathbf{j}) = \frac{\partial v}{\partial x} + \frac{\partial u}{\partial y} = 0.$$

$$\operatorname{curl}(v\mathbf{i} + u\mathbf{j}) = \begin{vmatrix} \mathbf{i} & \mathbf{j} & \mathbf{k} \\ \dfrac{\partial}{\partial x} & \dfrac{\partial}{\partial y} & 0 \\ v & u & 0 \end{vmatrix} = \mathbf{k}\left(\frac{\partial u}{\partial x} - \frac{\partial v}{\partial y}\right) = 0.$$

Hence

$$\frac{\partial u}{\partial x} = \frac{\partial v}{\partial y} \quad \text{and} \quad \frac{\partial v}{\partial x} = -\frac{\partial u}{\partial y}. \quad \text{(Cauchy–Riemann equations)}.$$

23.2 If $f(z) = u(x, y) + iv(x, y)$ is differentiable everywhere, and $u = v^2$, prove that $f(z)$ is constant.

Cauchy–Riemann equations $\partial u/\partial x = \partial v/\partial y$, $\partial v/\partial x = -\partial u/\partial y$ are satisfied since $f(z)$ is analytic. Now $u = v^2$ so $2v\,\partial v/\partial x = \partial u/\partial x = \partial v/\partial y$ and $\partial v/\partial x = -2v\,\partial v/\partial y$. Therefore

$$\frac{\partial v}{\partial y}(1 + 4v^2) = 0.$$

Hence either $v^2 = -\frac{1}{4}$ (which is not possible since v is real) or $\partial v/\partial y = 0$. In this case $v = g(x)$. Hence $\partial u/\partial x = 0$ and therefore $u = h(y)$, where h, g are arbitrary functions. For $h(y) = g^2(x)$ we must have both h and g constant functions. Therefore $f(z) = h + ig = \text{constant}$.

23.3 Evaluate $\int_{1+i}^{2+4i} z^2 \, dz$ (a) along the parabola $x = t$, $y = t^2$ where $1 \leqslant t \leqslant 2$, (b) along the straight line joining $1 + i$ and $2 + 4i$.

(a) $I = \int_{1+i}^{2+4i} z^2 \, dz. \qquad x = t, \ y = t^2, \qquad 1 \le t \le 2.$

$z = x + iy = t + it^2, \qquad z^2 = t^2 - t^4 + 2it^3, \qquad dz = dt(1 + 2it).$

Therefore

$I = \int_1^2 (t^2 - t^4 + 2it^3)(1 + 2it) \, dt = -\tfrac{86}{3} - 6i.$

(b) The equation of the line joining $(1, 1)$ and $(2, 4)$ is

$y = 3x - 2$

so

$z = x + iy = x + i(3x - 2) \quad \text{and} \quad dz = dx(1 + 3i).$

Hence

$I = \int_1^2 [-8x^2 + 12x - 4 + i(6x^2 - 4x)](1 + 3i) \, dx = -\tfrac{86}{3} - 6i.$

23.4   Use the Cauchy integral formula to evaluate

$I = \oint_C \dfrac{5z - 2}{z(z - 1)} \, dz,$

where C is (a) the circle $|z| = 2$, (b) the circle $|z| = \tfrac{1}{2}$.

Basic result

The Cauchy integral formula states that if $f(z)$ is an analytic function inside and on a contour C then

$f(z_0) = \dfrac{1}{2\pi i} \oint_C \dfrac{f(z)}{z - z_0} \, dz,$

where z_0 lies inside C. Now

$\oint_C \dfrac{5z - 2}{z(z - 1)} \, dz = \oint_C \dfrac{5z - 2}{z - 1} \, dz - \oint_C \dfrac{5z - 2}{z} \, dz$

by partial fractions. Using the integral formula each integral can be evaluated to give finally

$I = 2\pi i[(5z - 2)_{z=1} - (5z - 2)_{z=0}] = 10\pi i,$

since the poles of both integrands lie within $C: |z| = 2$.

When C is $|z| = \tfrac{1}{2}$, the pole of $\dfrac{5z - 2}{z - 1}$ is outside C and hence this function is analytic within C. Hence by the Cauchy theorem, which states that if $f(z)$ is any function analytic within and on C, then

$$\oint_C f(z)\,dz = 0, \text{ we have } \oint_C \frac{5z-2}{z-1}\,dz = 0.$$

The other integral takes the value $-4\pi i$ since $z=0$ is still within $|z|=\frac{1}{2}$. Hence in this case

$$I = \oint_C \frac{5z-2}{z(z-1)}\,dz = -(-4\pi i) = 4\pi i.$$

23.5 Find Laurent series for the following functions:

(a) $\dfrac{e^z}{z(z^2+1)}$ about $z=0$,   (b) $z\cos\dfrac{1}{z}$ about $z=0$,

(c) $\dfrac{e^z}{(z-1)^2}$ about $z=1$,   (d) $\dfrac{\cos^2 z}{z-\pi}$ about $z=\pi$.

(a) $e^z = 1 + z + \dfrac{z^2}{2!} + \dfrac{z^3}{3!} + \cdots.$

Hence

$$\frac{e^z}{z(z^2+1)} = \frac{1}{z}\left(1+z+\frac{z^2}{2!}+\frac{z^3}{3!}+\cdots\right)(1-z^2+z^4-z^6+\cdots),$$
$$\text{for } 0<|z|<1,$$

$$= \frac{1}{z}+1-\frac{z}{2}-\tfrac{5}{6}z^2+\cdots.$$

(b) $z\cos\dfrac{1}{z} = z\left(1-\dfrac{1}{2!}\dfrac{1}{z^2}+\dfrac{1}{4!}\dfrac{1}{z^4}-\cdots\right)$

$$= z - \frac{1}{2!}\frac{1}{z}+\frac{1}{4!}\frac{1}{z^3}-\cdots.$$

(c) In $e^z/(z-1)^2$ let $z-1=h$. Then

$$\frac{e^{1+h}}{h^2} = \frac{e}{h^2}\left(1+h+\frac{h^2}{2!}+\frac{h^3}{3!}+\cdots\right) = e\left(\frac{1}{h^2}+\frac{1}{h}+\frac{1}{2!}+\frac{h}{3!}+\cdots\right).$$

Hence

$$\frac{e^z}{(z-1)^2} = e\left(\frac{1}{(z-1)^2}+\frac{1}{(z-1)}+\frac{1}{2!}+\frac{(z-1)}{3!}+\cdots\right).$$

(d) In $\cos^2 z/(z - \pi)$ let $z - \pi = h$. Then

$$\frac{\cos^2 (h + \pi)}{h} = \frac{(\cosh \cos \pi - \sin h \sin \pi)^2}{h} = \frac{\cos^2 h}{h}$$

$$= \frac{1}{h} \left(1 - \frac{h^2}{2!} + \frac{h^4}{4!} - \frac{h^6}{6!} + \cdots \right)^2 = \frac{1}{h} - h + \frac{h^3}{3} - \cdots$$

Hence

$$\frac{\cos^2 z}{z - \pi} = \frac{1}{z - \pi} - (z - \pi) + \frac{(z - \pi)^3}{3!} - \cdots$$

23.6 Find the residues at the specified singular points of the following functions:

(a) $\dfrac{1}{1 - z}$ at $z = 1$,   (b) $\dfrac{e^{2z}}{z^4}$ at $z = 0$,

(c) $\dfrac{z^2 e^z}{(z - 3)^2}$ at $z = 3$,   (d) $\dfrac{1}{1 - e^z}$ at all poles.

(a) Residue at a simple pole at $z = a$ for a function $f(z)$ is given by

$$\lim_{z \to a} (z - a)f(z).$$

Hence in this case

$$\text{residue} = \lim_{z \to 1} (z - 1) \frac{1}{1 - z} = -1.$$

(b) Pole of order four at $z = 0$. The Laurent series is

$$\frac{1}{z^4} \left(1 + 2z + \frac{4z^2}{2!} + \frac{8z^3}{3!} + \cdots \right)$$

and the residue is the coefficient of the $1/z$ term. Hence residue $= \frac{4}{3}$.

(c) $\dfrac{z^2 e^z}{(z - 3)^2}$. Let $z - 3 = h$.

Then $\dfrac{(3 + h)^2 e^{3+h}}{h^2} = \dfrac{e^3}{h^2} \left(1 + h + \dfrac{h^2}{2!} + \ldots \right)(3 + h)^2$

$$= \frac{9e^3}{h^2} + \frac{15e^3}{h} + \frac{23}{2} + \cdots$$

Residue is coefficient of $1/h$ term. Hence residue $= 15e^3$.
Alternatively we can use the general result that for a pole of order

m at $z = a$ the residue is given by

$$\lim_{z \to a} \left\{ \frac{1}{(m-1)!} \frac{d^{m-1}}{dz^{m-1}} [(z-a)^m f(z)] \right\}.$$

In this case therefore (since at $z = 3$ the function has a pole of order two)

$$\text{residue} = \lim_{z \to 3} \left\{ \frac{d}{dz} \left((z-3)^2 \frac{z^2 e^z}{(z-3)^2} \right) \right\} = 15e^3.$$

(d) $\dfrac{1}{1-e^z}$. Poles occur at $e^z = 1$, which gives $z = 2k\pi i$, $k = 0, 1, 2, \ldots$, Put $z - 2\pi k i = h$. Then

$$\frac{1}{1-e^{(2\pi k i + h)}} = \frac{1}{1-e^h} = \frac{1}{1-\left(1+h+\dfrac{h^2}{2!}+\cdots\right)} = -\frac{1}{h\left(1+\dfrac{h}{2}+\cdots\right)}$$

$$= -\frac{1}{h}\left(1+\frac{h}{2}+\cdots\right)^{-1} = -\frac{1}{h}+\tfrac{1}{2}+\cdots.$$

Hence the residue is -1.

Alternatively for a simple pole at $z = a$, the residue at this point for a quotient function of the type $f(z)/g(z)$ is given by

$$\lim_{z \to a} \frac{f(z)}{g'(z)}.$$

In this case, since the poles at $z = 2\pi k i$ are all simple, we have (taking $f(z) = 1$, $g(z) = 1 - e^z$)

$$\text{residue} = \lim_{z \to 2\pi k i} \left(\frac{1}{-e^z} \right) = -\frac{1}{e^{2\pi k i}} = -1.$$

23.7 Show by integrating around a unit circle centred at the origin of the complex plane that

$$I = \int_0^{2\pi} \frac{\cos 3\theta}{5 - 4\cos\theta}\, d\theta = \frac{\pi}{12}.$$

In Fig. 41, C is the unit circle for which $|z| = 1$, $z = e^{i\theta}$. Now

$$\cos\theta = \frac{z+z^{-1}}{2}, \quad \cos 3\theta = \frac{z^3 + z^{-3}}{2},$$

$$dz = ie^{i\theta}\, d\theta = iz\, d\theta.$$

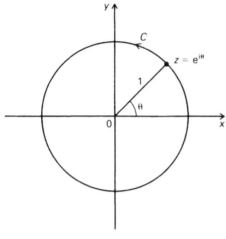

Fig. 41

Hence consider

$$\oint_C \frac{\frac{1}{2}(z^3 + z^{-3})}{5 - 4[(z + z^{-1})/2]} \frac{dz}{iz} = -\frac{1}{2i} \oint_C \frac{z^6 + 1}{z^3(2z - 1)(z - 2)} dz.$$

The integrand has a pole of order three at $z = 0$, a pole of order one (i.e. a simple pole) at $z = \frac{1}{2}$, and another pole of order one at $z = 2$, which lies *outside* C. Only the residues at $z = 0$ and at $z = \frac{1}{2}$ need be calculated.

Expanding

$$\frac{z^6 + 1}{z^3} \left[\frac{1}{(2z - 1)(z - 2)} \right]$$

as a Laurent series about $z = 0$, we have (by the binomial series)

$$\frac{1}{2} \left(\frac{z^6 + 1}{z^3} \right) (1 + 2z + 4z^2 + \cdots) \left(1 + \frac{z}{2} + \frac{z^2}{4} + \cdots \right).$$

The coefficient of $1/z$ is easily seen to be $\frac{21}{8}$, which is therefore the residue.

For the simple pole at $z = \frac{1}{2}$ we can use the alternative approach and evaluate

$$\text{residue} = \lim_{z \to 1/2} \left\{ (z - \tfrac{1}{2}) \frac{z^6 + 1}{z^3(2z - 1)(z - 2)} \right\} = -\frac{65}{24}.$$

Hence finally (using the Cauchy residue theorem)

$$I = -\frac{1}{2i} \oint_C = -\frac{1}{2i} 2\pi i (\tfrac{21}{8} - \tfrac{65}{24}) = \frac{\pi}{12}.$$

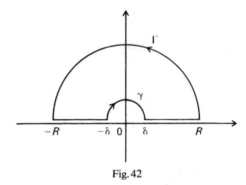

Fig. 42

23.8 Show, by integrating the function $(e^{ibz} - e^{iaz})/z^2$ around a semi-circular contour C in the upper half-plane and indented at the origin (see Fig. 42), that

$$\int_{-\infty}^{\infty} \frac{\cos bx - \cos ax}{x^2}\, dx = \pi(a - b),$$

where a and b are positive constants, $a \neq b$.

$$\frac{e^{ibz} - e^{iaz}}{z^2} = \frac{\left(1 + ibz - \dfrac{b^2 z^2}{2!} + \cdots\right) - \left(1 + iaz - \dfrac{a^2 z^2}{2!} + \cdots\right)}{z^2}$$

$$= \frac{i(b - a)}{z} - \frac{(b^2 - a^2)}{2!} + \frac{i(b^3 - a^3)}{3!} z + \cdots$$

Hence a simple pole at $z = 0$. Now

$$\oint_C \frac{e^{ibz} - e^{iaz}}{z^2}\, dz = 2\pi i \sum \text{residues within } C.$$

Therefore integrating around the contour shown, where Γ is the curved part of the semi-circle of radius R, and γ is the curved part of the semi-circle of radius ρ, we have

$$\oint_C = \int_{-R}^{-\delta} \frac{e^{ibx} - e^{iax}}{x^2}\, dx + \int_{\gamma} \frac{e^{ibz} - e^{iaz}}{z^2}\, dz + \int_{\delta}^{R} \frac{e^{ibx} - e^{iax}}{x^2}\, dx$$

$$+ \int_{\Gamma} \frac{e^{ibx} - e^{iaz}}{z^2}\, dz = 0.$$

Now

$$\left| \frac{e^{ibz} - e^{iaz}}{z^2} \right| \leq \frac{e^{-by} + e^{-ay}}{|z|^2}$$

(since $e^{ibz} = e^{ib(x+iy)} = e^{-by} e^{ibx}$ and hence $|e^{ibz}| = e^{-by}$). Since a, $b > 0$ and $y > 0$ as $|z| = R \to \infty$ this term behaves like $1/R^2$. Hence

$$\lim_{R \to \infty} \left| z \frac{e^{ibz} - e^{iaz}}{z^2} \right| = 0,$$

and therefore $\int_\Gamma \to 0$ as $R \to \infty$. On the small semi-circle γ, the function

$$\frac{e^{ibz} - e^{iaz}}{z^2} = \frac{i(b-a)}{z} + \text{Taylor series in } z \text{ (as already shown)}.$$

Hence

$$\int_\gamma = \int_\gamma \frac{i(b-a)}{z} \, dz + \int_\gamma \text{ Taylor series}.$$

Now on γ, $z = \rho e^{i\theta}$. Therefore

$$\int_\gamma = i(b-a) \int_\pi^0 \frac{i\rho e^{i\theta}}{\rho e^{i\theta}} \, d\theta + \text{terms in positive powers of } \rho.$$

As $\rho \to 0$, therefore,

$$\int_\gamma \to \pi(b-a).$$

Hence, remembering that $R \to \infty$ and $\rho \to 0$, we have

$$\int_{-\infty}^{\infty} \frac{e^{ibx} - e^{iax}}{x^2} \, dx + \pi(b-a) = 0,$$

(N.B. This integral must be interpreted as the principal value (see 5.8, 5.9) since there is a singularity in the integrand at $x = 0$).
    Taking the real part gives

$$\int_{-\infty}^{\infty} \frac{\cos bx - \cos ax}{x^2} \, dx = \pi(a-b).$$

23.9 Evaluate

$$\int_0^{\infty} \frac{\cos 3x}{(x^2+4)^2} \, dx$$

by integrating the function $e^{3iz}/(z^2+4)^2$ around the contour C defined by $z = Re^{i\theta}$, $0 \le \theta \le \pi$, and the real axis (see Fig. 43), and letting $R \to \infty$.

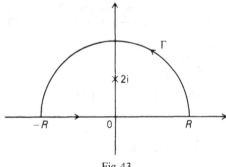

Fig. 43

The function $e^{3iz}/(z^2+4)^2$ has double poles at $z = \pm 2i$. Only the pole at $z = 2i$ lies within C. Now the residue at $z = 2i$ is

$$\lim_{z \to 2i} \left\{ \frac{d}{dz} \left((z-2i)^2 \frac{e^{3iz}}{(z^2+4)^2} \right) \right\} = \lim_{z \to 2i} \left\{ \frac{d}{dz} \left(\frac{e^{3iz}}{(z+2i)^2} \right) \right\} = -7ie^{-6}/32.$$

Also

$$\left| \frac{e^{3iz}}{(z^2+4)^2} \right| = \frac{\left| e^{3iz} \right|}{\left| z^2+4 \right|^2} = \frac{\left| e^{3iR(\cos\theta + i\sin\theta)} \right|}{\left| z^2+4 \right|^2} = \frac{e^{-3R\sin\theta}}{\left| z^2+4 \right|^2}$$

since $z = Re^{i\theta}$ on the curved part of C (called Γ, as in 23.8). For large R this term behaves like $e^{-3R\sin\theta}/R^4$ and therefore $\to 0$ as $R \to \infty$ for $0 \le \theta \le \pi$. Hence the integral along the curved part, Γ, of C tends to zero as $R \to \infty$. Hence finally, using the Cauchy residue theorem,

$$\int_{-\infty}^{\infty} + \int_{\Gamma} = \int_{-\infty}^{\infty} \frac{e^{3ix}}{(x^2+4)^2}\, dx + 0 = 2\pi i \left(-\frac{7ie^{-6}}{32} \right) = \tfrac{7}{16}e^{-6}.$$

Hence

$$\int_{-\infty}^{\infty} \frac{\cos 3x}{(x^2+4)^2}\, dx = \tfrac{7}{16}e^{-6} \quad \text{and therefore} \quad \int_{0}^{\infty} \frac{\cos 3x}{(x^2+4)^2}\, dx = \tfrac{7}{32}e^{-6}.$$

23.10 Evaluate $\displaystyle\int_{0}^{\infty} \frac{x^{1/4}}{1+x^2}\, dx$ by considering $\displaystyle\oint_{C} \frac{z^{1/4}}{1+z^2}\, dz$ around a suitably cut contour in the z-plane. (L.U.)

Using the same notation for the various parts of the contour as in the last two examples, we have (see Fig. 44.)

$$\oint_{C} \frac{z^{1/4}}{1+z^2}\, dz = \int_{\delta}^{R} \frac{z^{1/4}}{1+z^2}\, dx + \int_{\Gamma} \frac{z^{1/4}}{1+z^2}\, dz$$
$$+ \int_{R}^{\delta} \frac{z^{1/4}}{1+z^2}\, dx + \oint_{\gamma} \frac{z^{1/4}}{1+z^2}\, dz$$

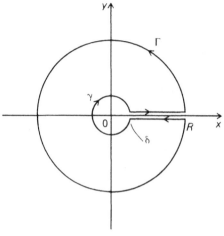

Fig. 44

The integrals around Γ and γ tend to zero as $R \to \infty$ and $\rho \to 0$ respectively. In the first integral on the right-hand side (representing the integral along the upper straight path) $z = xe^{i0} = x$. In the third integral on the right-hand side the correct argument of z must be taken since we are taking a fractional power. Accordingly along the *lower* straight path z should be written as $xe^{2\pi i}$, since we have increased the argument by 2π. Hence

$$\int_R^\delta \frac{(xe^{2\pi i})^{1/4}\,dx}{(xe^{2\pi i})^2+1} = e^{\pi i/2}\int_R^\delta \frac{x^{1/4}}{1+x^2}\,dx.$$

In the limit as $R \to \infty$ and $\rho \to 0$ therefore

$$\lim_{\substack{R \to \infty \\ \rho \to 0}} \left\{ (1-e^{i\pi/2})\int_\delta^R \frac{x^{1/4}}{1+x^2}\,dx \right\} = 2\pi i \sum \text{ residues at poles within } C.$$

Now the poles occur where $z^2+1 = 0$, i.e. at $z = \pm i$. Residue at $z = i$, $(=e^{i\pi/2})$ is

$$\lim_{z \to i} \left\{ (z-i)\frac{z^{1/4}}{z^2+1} \right\} = \lim_{z \to i} \left\{ \frac{z^{1/4}}{z+i} \right\} = \frac{e^{i\pi/8}}{2i}.$$

Similarly the residue at $z = -i (=e^{3\pi i/2})$ is

$$\lim_{z \to -i} \left\{ (z-(-i))\frac{z^{1/4}}{z^2+1} \right\} = -\frac{1}{2i}e^{3\pi i/8}.$$

Hence

$$\oint_C = 2\pi i\frac{1}{2i}(e^{i\pi/8} - e^{3\pi i/8}) = (1-e^{i\pi/2})\int_0^\infty \frac{x^{1/4}}{1+x^2}\,dx,$$

and therefore

$$\int_0^\infty \frac{x^{1/4}}{1+x^2}\,dx = \frac{\pi}{1-i}\,e^{i\pi/4}(e^{-i\pi/8}-e^{i\pi/8}) = \frac{\pi(1+i)}{2}\,e^{i\pi/4}\left(-2i\sin\frac{\pi}{8}\right)$$

$$= \sqrt{2}\pi\,\sin\frac{\pi}{8}.$$

23.11 Define $\log_e z$ and show that for $x>0$ the principal value of $\log(-x)$ is $\log_e|x|+\pi i$. Show that $z=0$ is a branch point of $\log_e z$. By integrating $\log_e z/(z^2+a^2)$ around a semi-circle in the upper half plane, centred and indented at $z=0$, evaluate

$$\int_0^\infty \frac{\log_e x}{x^2+a^2}\,dx,$$

where $a>0$. (L.U.)

$$\log_e z = \log_e re^{i(\theta+2\pi n)} = \log_e r + i(\theta+2\pi n)$$
$$= \log_e|z|+i(\theta+2\pi n), \qquad n=0,1,2,3,\ldots.$$

Let $z=xe^{\pi i}=-x$. Then $\log_e(-x)=\log_e|x|+i\pi$ is the principal value. The branch point of $\log_e z$ is at $z=0$ since $\log_e|z|$ increases by $2\pi i$ as z describes any circle $z=re^{i\theta}$ in the positive direction (anti-clockwise) about 0.

Consider Fig. 45. Then

$$\oint_C \frac{\log_e z}{z^2+a^2}\,dz = \int_{-R}^{-\delta} \frac{\log_e(xe^{\pi i})}{(xe^{\pi i})+a^2} + \int_\gamma \cdots + \int_\delta^R \frac{\log_e x}{x^2+a^2}\,dx + \int_\Gamma \cdots$$

The only pole inside C is at $z=ia$.

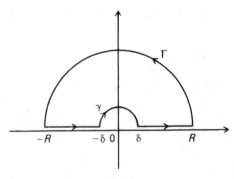

Fig. 45

The residue at $z = ia$ is

$$\lim_{z \to ia} \left\{ (z - ia) \frac{\log_e z}{(z - ia)(z + ia)} \right\} = \frac{1}{2ia} \log_e ia$$

$$= \frac{1}{2ia} \left[\log_e a + \frac{i\pi}{2} \right] = \frac{\pi}{4a} + \frac{\log_e a}{2ia}.$$

On Γ: $\int_\Gamma \to 0$ as $R \to \infty$ since the integrand behaves like $(\log_e R)/R^2$. On γ: $\int_\gamma \to 0$ as $\rho \to 0$ since

$$\lim_{\rho \to 0} \left[\frac{\rho \log_e \rho}{\rho^2 + a^2} \right] = 0.$$

Hence as $R \to \infty$, $\rho \to 0$ ($\delta \to 0$)

$$\lim_{\substack{R \to \infty \\ \rho \to 0}} \left[\int_{-R}^{-\delta} \frac{\log_e (xe^{\pi i})}{(xe^{\pi i})^2 + a^2} dx + \int_\delta^R \frac{\log_e x}{x^2 + a^2} dx \right]$$

$$= 2\pi i \left[\frac{\pi}{4a} + \frac{1}{2ia} \log_e a \right].$$

Therefore since $\log_e (xe^{\pi i}) = \log_e (-x) = \log_e |x| + i\pi$,

$$2 \int_0^\infty \frac{\log_e x}{x^2 + a^2} dx + \frac{i\pi}{a} \int_0^\infty \frac{a \, dx}{x^2 + a^2} = \frac{\pi^2 i}{2a} + \frac{\pi}{a} \log_e a,$$

or $\left(\text{since } \int_0^\infty \frac{a \, dx}{a^2 + x^2} = \frac{\pi}{2} \right)$

$$\int_0^\infty \frac{\log_e x}{x^2 + a^2} dx = \frac{\pi}{2a} \log_e a.$$

23.12 Show, by contour integration, that

$$\int_0^\infty \frac{x^3 \, dx}{(e^x - 1)} = \frac{\pi^4}{15}.$$

Consider the indented contour C shown in Fig. 46. where

$$0 \leqslant x \leqslant R, \ y = 0; \ 0 \leqslant y \leqslant 2\pi, \ x = R;$$
$$R \geqslant x \geqslant r, \ y = 2\pi; \ 2\pi - r \geqslant y \geqslant 0, \ x = 0.$$

If the complex function were $z^3/(e^z - 1)$, the bottom would have the same form $e^x - 1$ on both $y = 0$ and $y = 2\pi$, which is the reason for choosing the above contour. However, the numerators, x^3 and $-(x + 2\pi i)^3$, in the sum would cancel at order x^3. We therefore consider the function

$$f(z) = (\alpha z^5 + \beta z^3)/(e^z - 1),$$

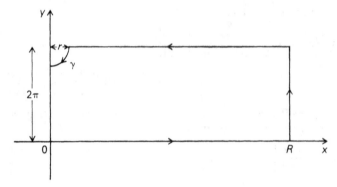

Fig. 46

where α and β have to be chosen to give a sum $x^3/(e^x - 1)$ from the two horizontal parts of the contour. It is easily found that $\alpha = \dfrac{1}{40\pi^2}$ and $\beta = \frac{1}{6}$. Now $(e^z - 1)$ has zeros at $z = 2n\pi i$, where $n = 0, \pm 1, \pm 2, \ldots$. However, $z = 0$ is not a pole of $f(z)$ since $(e^z - 1)$ itself behaves like z for small z. The pole at $z = 2\pi i$ is avoided by the small circular quadrant γ of radius r. There are no singularities within C. Hence by Cauchy's theorem

$$\oint_C \frac{(1/40\pi^2)z^5 + \frac{1}{6}z^3}{e^z - 1}\, dz = 0.$$

This gives

$$0 = \int_0^R \frac{(1/40\pi^2)x^5 + \frac{1}{6}x^3}{e^x - 1}\, dx$$

$$+ \int_0^{2\pi} \frac{[(1/40\pi^2)(R+iy)^5 + \frac{1}{6}(R+iy)^3]}{e^{R+iy} - 1}\, i\, dy$$

$$+ \int_R^0 \frac{(1/40\pi^2)(x+2\pi i)^5 + \frac{1}{6}(x+2\pi i)^3}{e^x - 1}\, dx + \int_\gamma f(z)\, dz$$

$$+ \int_{2\pi - r}^0 \frac{(1/40\pi^2)(iy)^5 + \frac{1}{6}(iy)^3}{(e^{iy} - 1)}\, i\, dy.$$

From this follows

$$\int_0^R \frac{(x^3 + \text{imag. terms})}{(e^x - 1)}\, dx + \int_\gamma f(z)\, dz$$

$$+ \int_0^{2\pi - r} \left[\left(\frac{y^5}{40\pi^2} - \frac{y^3}{6} \right) \frac{(\cos y - 1 - i\sin y)}{2(1 - \cos y)} \right] dy$$

where we have used $R^n e^{-R} \to 0$ as $R \to \infty$ to eliminate the integral over $x = R$, $0 \leqslant y \leqslant 2\pi$.

218

Now putting $z = 2\pi i + re^{i\theta}$ on γ and letting $r \to 0$ (with $e^z - 1 = e^{2\pi i + re^{i\theta}} - 1 = re^{i\theta} + $ higher powers in r) we have

$$\int_\gamma \frac{(1/40\pi^2)z^5 + \frac{1}{6}z^3}{(e^z - 1)}\, dz = \int_0^{-\pi/2} \frac{(2\pi i)^5 / (40\pi^2) + \dfrac{(2\pi i)^3}{6}}{re^{i\theta}}\, ire^{i\theta}\, d\theta$$

$$= \left(-\frac{4\pi^3}{5} + \frac{4\pi^3}{3}\right)\left(-\frac{\pi}{2}\right) = -\frac{4\pi^4}{15}.$$

Inserting into the previous equation and taking the *real* part (with $R \to \infty$, $r \to 0$) we finally obtain

$$\int_0^\infty \frac{x^3\, dx}{(e^x - 1)} - \frac{4\pi^4}{15} + \int_0^{2\pi}\left(\frac{y^5}{40\pi^2} - \frac{y^3}{6}\right)\frac{dy}{(-2)} = 0,$$

whence

$$\int_0^\infty \frac{x^3\, dx}{(e^x - 1)} - \frac{4\pi^4}{15} - \frac{\pi^4}{2}(\tfrac{4}{15} - \tfrac{2}{3}) = 0$$

or

$$\int_0^\infty \frac{x^3\, dx}{(e^x - 1)} = \frac{\pi^4}{15}.$$

(Problem and solution contributed by Professor Trevor Stuart).

24. FOURIER TRANSFORMS

[Some standard transforms, inversion theorem, sine and cosine transforms, convolution theorem, simple partial differential equations]

24.1 (a) Show that the Fourier transform of

$$f(x) = \begin{cases} 1, & |x| < a \\ 0, & |x| > a \end{cases}$$

is $2\dfrac{\sin as}{s}$.

(b) Using the result

$$\int_0^\infty e^{-\lambda x^2} \cos sx \, dx = \left(\frac{\pi}{4\lambda}\right)^{1/2} e^{-s^2/4\lambda},$$

obtain the Fourier transform of $e^{-\lambda x^2}$, and hence derive the Fourier transform of the delta function $\delta(x)$ by a limiting process using the definition

$$\delta(x) = \lim_{n \to \infty} \sqrt{\frac{n}{\pi}} \, e^{-nx^2}.$$

Basic result

The Fourier transform of $f(x)$, $F\{f(x)\}$, is defined as $\bar{f}(s) = \int_{-\infty}^\infty f(x) e^{isx} \, ds$.

(a) $\displaystyle \bar{f}(s) = \int_{-\infty}^{-a} 0 \cdot e^{isx} \, dx + \int_{-a}^a 1 \cdot e^{isx} \, dx + \int_a^\infty 0 \cdot e^{isx} \, dx$

$\displaystyle \qquad = \frac{1}{is}(e^{isa} - e^{-isa}) = \frac{2 \sin as}{s}.$

(b) $\displaystyle \int_{-\infty}^\infty e^{-\lambda x^2} e^{isx} \, dx = 2\int_0^\infty e^{-\lambda x^2} \cos sx \, dx \left(\text{since } \cos sx = \frac{e^{isx} + e^{-isx}}{2}\right)$

$\displaystyle \qquad = (\pi/\lambda)^{1/2} e^{-s^2/4\lambda}, \quad \text{(using the given integral.)}$

Now

$$F\left\{\sqrt{\frac{n}{\pi}} e^{-nx^2}\right\} = \sqrt{\frac{n}{\pi}} \sqrt{\frac{\pi}{n}} e^{-s^2/4n} = e^{-s^2/4n}$$

so if

$$\delta(x) = \lim_{n \to \infty} \sqrt{\frac{n}{\pi}} e^{-nx^2}, \quad \text{then} \quad F\{\delta(x)\} = \lim_{n \to \infty} e^{-s^2/4n} = 1.$$

(See 18.9. for $\delta(x)$).

24.2 Find the Fourier transform of

$$f(x) = e^{-a|x|} \quad (a \text{ real and positive})$$

and find the transform of $x^n e^{-a|x|}$ (n integral) in terms of a derivative.

Now

$$F\{e^{-a|x|}\} = \bar{f}(s) = \int_{-\infty}^{\infty} e^{-a|x|}e^{isx}\, dx$$

$$= \int_{-\infty}^{0} e^{ax}e^{isx}\, dx + \int_{0}^{\infty} e^{-ax}e^{isx}\, dx$$

$$= \left(\frac{e^{(is-a)x}}{is-a}\right)_{0}^{\infty} + \left(\frac{e^{(is+a)x}}{is+a}\right)_{-\infty}^{0} = \frac{2a}{a^2+s^2}.$$

$$F\{x^n e^{-a|x|}\} = \int_{-\infty}^{\infty} x^n\, e^{-a|x|}\, e^{isx}\, dx$$

$$= \int_{-\infty}^{\infty} \left(\frac{\partial}{\partial s}\right)^n \left(\frac{e^{-a|x|}e^{isx}}{i^n}\right) dx$$

$$= (-i)^n \frac{d^n}{ds^n} \int_{-\infty}^{\infty} e^{-a|x|}e^{isx}\, dx$$

$$= (-i)^n \frac{d^n}{ds^n}\left(\frac{2a}{a^2+s^2}\right).$$

24.3 Given that $F\{e^{-|x|}\} = 2/(1+s^2)$, use the inversion theorem to prove that

$$\int_{-\infty}^{\infty} \frac{\cos x}{1+x^2}\, dx = \frac{\pi}{e}.$$

Basic result

The inversion theorem states that if $F\{f(x)\} = \bar{f}(s)$ is the Fourier transform of $f(x)$, then

$$f(x) = \frac{1}{2\pi}\int_{-\infty}^{\infty} \bar{f}(s)e^{-isx}\, ds.$$

By the inversion theorem

$$e^{-|x|} = \frac{1}{2\pi}\int_{-\infty}^{\infty} \frac{2}{1+s^2}e^{-isx}\, ds \quad \text{or} \quad \int_{-\infty}^{\infty} \frac{e^{-isx}}{1+s^2}\, ds = \pi e^{-|x|}.$$

Since

$$\int_{-\infty}^{\infty} \frac{e^{-isx}}{1+s^2}\, ds = \int_{-\infty}^{\infty} \frac{\cos sx + i\sin sx}{1+s^2}\, ds,$$

we find

$$\int_{-\infty}^{\infty} \frac{\cos sx}{1+s^2} \, ds = \pi e^{-|x|}.$$

Hence with $x = 1$,

$$\int_{-\infty}^{\infty} \frac{\cos s}{1+s^2} \, ds \left(= \int_{-\infty}^{\infty} \frac{\cos x}{1+x^2} \, dx \right) = \frac{\pi}{e}.$$

24.4 If $F\{f(x)\} = \bar{f}(s)$, show that

(a) $F\{f(x+c)\} = e^{isc}\bar{f}(s)$ (c constant),   (b) $F\{e^{cx}f(x)\} = \bar{f}(s-ic)$,

(c) $F\{xf(x)\} = -i \dfrac{d}{ds} \bar{f}(s)$.

(d) $F\{f'(x)\} = -is\bar{f}(s)$.

(a) $F\{f(x+c)\} = \displaystyle\int_{-\infty}^{\infty} f(x+c)e^{isx} \, dx$

$$= \int_{-\infty}^{\infty} f(u)e^{is(u-c)} \, du \qquad (u = x+c)$$

$$= e^{-isc}\int_{-\infty}^{\infty} f(u)e^{isu} \, du = e^{-isc}\bar{f}(s).$$

(b) $F\{e^{cx}f(x)\} = \displaystyle\int_{-\infty}^{\infty} e^{cx}f(x)e^{isx} \, dx = \int_{-\infty}^{\infty} e^{i(s-ic)x}f(x) \, dx$

$$= \bar{f}(s-ic).$$

(c) $F(xf(x)) = \displaystyle\int_{-\infty}^{\infty} xf(x)e^{isx} \, dx = -i\int_{-\infty}^{\infty} \frac{\partial}{\partial s}(f(x)e^{isx}) \, dx$

$$= -i\frac{d}{ds}\int_{-\infty}^{\infty} f(x)e^{isx} \, dx = -i\frac{d}{ds}\bar{f}(s).$$

(d) $F\{f'(x)\} = \displaystyle\int_{-\infty}^{\infty} f'(x)e^{isx} \, dx = (f(x)e^{isx})_{-\infty}^{\infty} - is\int_{-\infty}^{\infty} f(x)e^{isx} \, dx$

(integrating by parts). Assuming $f(x) \to 0$ as $x \to \pm\infty$

$$F\{f'(x)\} = -is\bar{f}(s).$$

24.5 Find the Fourier sine and cosine transforms of e^{-ax} and $1/x$, where a is a constant.

Fourier sine transform:

$$F_s\{e^{-ax}\} = \int_0^\infty e^{-ax}\sin sx\,dx = \frac{s}{a^2+s^2}.$$

Fourier cosine transform:

$$F_c\{e^{-ax}\} = \int_0^\infty e^{-ax}\cos sx\,dx = \frac{a}{a^2+s^2}.$$

$$F_s\left\{\frac{1}{x}\right\} = \int_0^\infty \frac{1}{x}\sin sx\,dx = \frac{\pi}{2}$$

using the standard integral (see 18.12)

$$\int_0^\infty \frac{\sin x}{x}\,dx = \frac{\pi}{2}.$$

$$F_c\left\{\frac{1}{x}\right\} = \int_0^\infty \frac{1}{x}\cos sx\,dx$$

does not exist since the integral is divergent.

24.6 Show by differentiation under the integral sign that if s is real, and

$$F\{e^{-x^2}\} = \int_{-\infty}^\infty e^{-x^2}e^{isx}\,dx = \bar{f}(s),$$

where $\bar{f}(s)$ is the Fourier transform of $f(x) = e^{-x^2}$,

$$\frac{d\bar{f}(s)}{ds} + \tfrac{1}{2}s\bar{f}(s) = 0.$$

Hence show that $\bar{f}(s) = \sqrt{\pi}\,e^{-s^2/4}$ given $\bar{f}(0) = \sqrt{\pi}$.

Verify that $F\{e^{-c|x|}\} = \frac{2c}{c^2+s^2}$, where c is a positive constant, and use the Fourier inversion theorem to deduce that

$$F\left\{\frac{1}{c^2+x^2}\right\} = \frac{\pi}{c}e^{-c|s|}.$$

Hence using the convolution theorem

$$F\left\{\int_{-\infty}^\infty f(t)g(x-t)\,dt\right\} = \bar{f}(s)\bar{g}(s)$$

deduce the Fourier transform of

$$\int_{-\infty}^\infty \frac{e^{-(x-t)^2}}{c^2+t^2}\,dt.$$

$$\bar{f}(s) = \int_{-\infty}^{\infty} e^{-x^2} e^{isx} \, dx.$$

$$\frac{d\bar{f}(s)}{ds} = i \int_{-\infty}^{\infty} x e^{-x^2} e^{isx} \, dx$$

$$= i \left[\left(\frac{e^{isx} e^{-x^2}}{-2} \right)_{-\infty}^{\infty} + \frac{is}{2} \int_{-\infty}^{\infty} e^{-x^2} e^{isx} \, dx \right] = -\frac{s}{2} \bar{f}(s).$$

Integrating gives $\log_e \bar{f}(s) = -s^2/4 + A$, where A is an integration constant, whence $\bar{f}(s) = \bar{f}(0) e^{-s^2/4} = \sqrt{\pi} e^{-s^2/4}$. If $f(x) = e^{-c|x|}$, then

$$\bar{f}(s) = \int_{-\infty}^{\infty} e^{-c|x|} e^{isx} \, dx$$

$$= \int_{-\infty}^{0} e^{cx} e^{isx} \, dx + \int_{0}^{\infty} e^{-cx} e^{isx} \, dx$$

$$= \frac{1}{c - is} + \frac{1}{c + is} = \frac{2c}{c^2 + s^2}.$$

By the inversion theorem the Fourier transform of $2c/(c^2 + x^2)$ is $2\pi e^{-c|s|}$. Hence

$$F\left\{ \frac{1}{c^2 + x^2} \right\} = \frac{\pi}{c} e^{-c|s|}.$$

In the convolution theorem let

$$f(t) = \frac{1}{c^2 + t^2}, \quad g(t) = e^{-t^2}.$$

Then

$$F\left\{ \int_{-\infty}^{\infty} \frac{1}{c^2 + t^2} \cdot e^{-(x-t)^2} \, dt \right\} = \bar{f}(s)\bar{g}(s) = \frac{\pi}{c} e^{-c|s|} \sqrt{\pi} e^{-s^2/4}$$

$$= \frac{\pi^{3/2}}{c} e^{-c|s| - s^2/4}.$$

24.7 Solve the equation $\dfrac{\partial^2 u}{\partial x^2} = \dfrac{1}{k} \dfrac{\partial u}{\partial t}$ given $u(0, t) = u_0$, where u_0 is a constant, and $u(x, 0) = 0$, using the Fourier sine transformation.

It can easily be shown from the Fourier sine transform that

$$\int_{0}^{\infty} \frac{\partial^2 u}{\partial x^2} \sin sx \, dx = -s^2 \bar{u}(s, t) + su(0, t),$$

where $\bar{u}(s, t)$ is the Fourier sine transform of $u(x, t)$ with respect to x.

Hence the equation when transformed becomes

$$-s^2\bar{u}(s, t) + su(0, t) = \frac{1}{k}\frac{d}{dt}\bar{u}(s, t).$$

Inserting the boundary condition $u(0, t) = u_0$ we have

$$\frac{d}{dt}\bar{u}(s, t) + ks^2\bar{u}(s, t) = ksu_0.$$

The solution of this equation is

$$\bar{u}(s, t)e^{ks^2t} = ksu_0\int e^{ks^2t}\,dt + A,$$

or

$$\bar{u}(s, t) = Ae^{-ks^2t} + \frac{u_0}{s},$$

where A is an integration constant.
Now since $u(x, 0) = 0$,

$$\bar{u}(s, 0) = \int_0^\infty u(x, 0)\sin sx\,dx = 0.$$

Hence

$$A = -\frac{u_0}{s} \quad \text{and} \quad \bar{u}(s, t) = \frac{u_0}{s}(1 - e^{-ks^2t}).$$

Using the inversion formula for the Fourier sine transform

$$u(x, t) = \frac{2u_0}{\pi}\int_0^\infty (1 - e^{-ks^2t})\frac{\sin sx}{s}\,ds$$

$$= u_0\left(1 - \frac{2}{\pi}\int_0^\infty e^{-ks^2t}\frac{\sin sx}{s}\,ds\right)$$

since

$$\int_0^\infty \frac{\sin sx}{s}\,ds = \frac{\pi}{2} \quad (x > 0)\,(\text{see } 18.12).$$

Expanding $\sin sx$ in series and integrating term-by-term we have finally

$$u(x, t) = u_0\left(1 - \operatorname{erf}\frac{x}{2\sqrt{kt}}\right)$$

(see 5.11 for properties of erf x).

25. CALCULUS OF VARIATIONS

[Euler equation, extremal curves, discontinuous paths, approximate solution of differential equations]

25.1 Obtain the stationary value of the integral

$$I = \int_A^B (y'^2 + 2xy - y^2)\,dx,$$

where A and B are the points $(0, 0)$ and $(\pi/2, \pi/2)$ respectively in
the xy-plane.

Basic result

For stationary value of the integral $\int_A^B f(x, y, y')\,dx$, where $f(x, y, y')$
is a differentiable function and $y' = dy/dx$, we have to solve the
Euler equation

$$\frac{\partial f}{\partial y} - \frac{d}{dx}\left(\frac{\partial f}{\partial y'}\right) = 0$$

for $y(x)$. This gives the extremal curve, which when substituted back
into the integral gives its stationary value.

Here $f = y'^2 + 2xy - y^2$. Hence the Euler equation becomes

$$(2x - 2y) - \frac{d}{dx}(2y') = 0$$

or

$$\frac{d^2y}{dx^2} + y = x.$$

Hence $y = C \cos x + D \sin x + x$, where C and D are integration
constants. Since the end points have to lie on this curve we find
$C = D = 0$. Hence $y = x$ is the extremal curve. Inserting back into
the integral and performing the integration gives the stationary

value

$$I = \int_0^{\pi/2} (1^2 + 2x^2 - x^2)\, dx = \frac{\pi}{2}\left(1 + \frac{\pi^2}{12}\right).$$

25.2 Find the solution of the Euler equation for the integral

$$I = \int_A^B \left\{ \left(\frac{dy}{dx}\right)^2 \left(1 + \frac{dy}{dx}\right)^2 \right\} dx,$$

where $A = (0,0)$ and $B = (1,2)$ in the xy-plane.

The integrand $f = y'^2(1 + y')^2$ is explicitly independent of y. Hence the Euler equation is

$$\frac{d}{dx}\left(\frac{\partial f}{\partial y'}\right) = 0$$

or, integrating, $\partial f/\partial y' = C$, where C is an integration constant. Now

$$\frac{\partial f}{\partial y'} = 2y'^2(1 + y') + 2y'(1 + y')^2 = C.$$

Therefore $y' = $ constant and hence $y = \alpha x + \beta$, where α and β are integration constants. Inserting the end-point values, we find $\alpha = 2$, $\beta = 0$, and therefore $y = 2x$ is the extremal curve.

25.3 Find the extremal curves of the integral

$$I = \int_A^B \frac{y'^2}{1 + y^2}\, dx,$$

where $A = (0,0)$, $B = (1,2)$.

Here $f = \dfrac{y'^2}{1 + y^2}$ is *explicitly* independent of x. The Euler equation may be integrated in this case (basic result) to give

$$f - y' \frac{\partial f}{\partial y'} = C,$$

where C is a constant. Hence

$$\frac{y'^2}{1 + y^2} - y'\left(\frac{2y'}{1 + y^2}\right) = C, \quad \text{or} \quad \frac{y'^2}{1 + y^2} = -C (=\alpha^2, \text{say})$$

and therefore $y' = \alpha\sqrt{1 + y^2}$.

Integrating we have $y = \sinh(\alpha x + \beta)$ and the end-points require that the integration constants are $\beta = 0$ and $\alpha = \sinh^{-1} 2$. The solution is therefore

$$y = \sinh(x \sinh^{-1} 2).$$

25.4 Find the function $y(x)$ which makes the integral

$$I = \int_0^{\pi/2} (y'^2 + 2xyy') \, dx,$$

with $y(0) = 0$, $y(\pi/2) = 1$ stationary, where $y(x)$ is subject to the constraint

$$\int_0^{\pi/2} y \, dx = \frac{\pi}{2} - 1.$$

Consider $f + \lambda g$, where f is the integrand of I and g is the integrand of the constraint integral, and λ is a parameter. We now consider the Euler equation for the function $F = f + \lambda g$ (the Lagrange multiplier method). Hence, since $F = y'^2 + 2xyy' + \lambda y$, the Euler equation is

$$2xy' + \lambda = 2y'' + 2y + 2xy'.$$

Hence

$$y'' + y = \lambda/2$$

and $y = (\lambda/2) + A \cos x + B \sin x$, where A and B are integration constants. Now insert this form of y into the constraint integral

$$\int_0^{\pi/2} y \, dx = \int_0^{\pi/2} \left(\frac{\lambda}{2} + A \cos x + B \sin x\right) dx = \frac{\pi\lambda}{4} + A + B = \frac{\pi}{2} - 1,$$

whence

$$A + B = \frac{\pi}{2} - \frac{\pi\lambda}{4} - 1.$$

Also the end-point conditions applied to the solution for y give

$$0 = \frac{\lambda}{2} + A \quad \text{and} \quad 1 = \frac{\lambda}{2} + B.$$

Hence $A = -\frac{\lambda}{2}$, $B = 1 - \frac{\lambda}{2}$ and therefore $1 - \lambda = \frac{\pi}{2} - \frac{\pi\lambda}{4} - 1$, whence $\lambda = 2$. Accordingly $A = -1$, $B = 0$ and $y = 1 - \cos x$.

25.5 The performance index, I, of a system is given by

$$I = \int_0^1 f(\dot{y})g(y) \, dt,$$

where the control variable $y(t)$ has fixed values at $t = 0$ and $t = 1$ (g

is a *non-constant* function of y, and $\dot{y} = dy/dt$). Using the Euler equation, show that if at an extremum of I,

$$f(\dot{y})g(y) = k,$$

where k is a constant, then

$$\dot{y} = \frac{ff'}{f'^2 - ff''},$$

where $f' = df/d\dot{y}$, $f'' = d^2f/d\dot{y}^2$. Verify that, for $0 < t < 1$, $f = A\dot{y}^n$, where A and n are constants, satisfies the equation.

Given that $g(y) = e^{2y}$ and $n = 2$, determine $y(t)$ given $y(0) = 0$ and $y(1) = 1$.   (L.U.)

Euler equation for $f(\dot{y})g(y)$ is

$$\frac{\partial}{\partial y}(f(\dot{y})g(y)) - \frac{d}{dt}\left[\frac{\partial}{\partial \dot{y}}(f(\dot{y})g(y))\right] = 0.$$

Therefore $f(\dot{y})g'(y) - f''(\dot{y})\ddot{y}g(y) - f'(\dot{y})g'(y)\dot{y} = 0$ (dashes denoting differentiation with respect to the argument of the function). Now since $f(\dot{y})g(y) = k$, we have, by differentiating with respect to t,

$$f'(\dot{y})\ddot{y}g(y) + f(\dot{y})g'(y)\dot{y} = 0.$$

Hence

$$\ddot{y} = -\frac{f(\dot{y})g'(y)\dot{y}}{f'(\dot{y})g(y)}.$$

Inserting into the Euler equation gives

$$\dot{y} = \frac{ff'}{f'^2 - ff''}$$

Putting $f = A\dot{y}^n$ into this expression for $\dot{y}$ gives

$$\dot{y} = \frac{A\dot{y}^n n A\dot{y}^{n-1}}{(An\dot{y}^{n-1})^2 - A\dot{y}^n An(n-1)\dot{y}^{n-2}}$$

$$= \frac{A^2 n\dot{y}^{2n-1}}{A^2\dot{y}^{2n-2}[n^2 - n(n-1)]} = \dot{y},$$

so the equation is satisfied. Now $f(\dot{y})g(y) = k$ so $A\dot{y}^n g(y) = k$, or

$$\dot{y}^n = \frac{\alpha^2}{g(y)} \qquad \left(\alpha^2 = \frac{k}{A}\right).$$

If $n = 2$, $\dot{y}^2 = \alpha^2/e^{2y}$. Therefore, integrating, we have $e^y = \alpha t + \beta$, where β is an integration constant). The end conditions are $y(0) = 0$, $y(1) = 1$, so $\alpha = e - 1$, $\beta = 1$. Hence finally $y = \log_e(1 + (e-1)t)$.

25.6 Obtain the extremal curves for

$$I = \int_A^B x^2 y'^2 \, dx$$

in the two cases (a) $A = (1, 2)$, $B = (2, 1)$. (b) $A = (1, 2)$, $B = (-2, -1)$.

The Euler equation is

$$\frac{d}{dx}(2x^2 y') = 0.$$

Therefore $y' = c/2x^2$, and hence $y = \alpha/x + \beta$, where α and β are integration constants.
(a) Using the end-point values we find $\alpha = 2$, $\beta = 0$. Hence the extremal curve is the rectangular hyperbola $xy = 2$ connecting $A(1, 2)$ and $B(2, 1)$ (see Fig. 47).
(b) From the solution $y = \alpha/x + \beta$ it is easily seen that no α and β values can be found which allow the points $(1, 2)$, $(-2, -1)$ to lie on the *smooth* extremal curve. However, if we allow extremal curves with *discontinuous derivatives*, then we can obtain an extremal curve joining A and B. This is possible since the integrand is a perfect square and its minimum value is when $y' = 0$ (i.e. along PQ). No contribution to the integral comes from AQ and PB since in both cases $x = $ constant, $dx = 0$ (see Fig. 47). An extremal path is therefore AQPB.

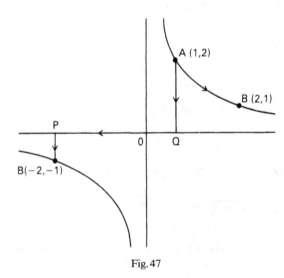

Fig. 47

25.7 Show that the differential equation

$$\frac{d^2y}{dx^2} + xy = x$$

is the Euler equation for the integral

$$I = \int_A^B (2xy - xy^2 + y'^2)\,dx.$$

Taking a trial function $y = ax(1-x)$ as an approximate solution of the differential equation which vanishes at $x = 0$ and $x = 1$, minimize I with respect to a and show that $a = -\frac{5}{19}$.

$$\frac{\partial f}{\partial y} = 2x - 2xy, \frac{\partial f}{\partial y'} = 2y'.$$

Hence the Euler equation is $2x - 2xy - 2y'' = 0$ or $y'' + xy = x$. Putting $y = ax(1-x)$ into the integral we have

$$I = \int_0^1 [2ax^2(1-x) - a^2x^3(1-x)^2 + a^2(1-2x)^2]\,dx = \frac{a}{6} + \frac{19a^2}{60}.$$

Therefore

$$\frac{dI}{da} = \frac{1}{6} + \frac{19a}{30} = 0,$$

whence $a = -\frac{5}{19}$. In this way we obtain an approximate solution of the differential equation. This approach forms the basis of the Rayleigh–Ritz (or direct) method.

---

26. SUFFIX NOTATION

[Summation convention, vector operators, δ_{ij} and ε_{ijk} symbols]

26.1 Write in expanded form $a_{ik}x_ix_k$, where $a_{ik} = a_{ki}(i, k = 1, 2, 3)$.

Basic result

Unless otherwise stated, repeated indices are summed (Einstein summation convention).

$$a_{ik}x_ix_k = a_{11}x_1x_1 + a_{12}x_1x_2 + a_{13}x_1x_3$$
$$+ a_{21}x_2x_1 + a_{22}x_2x_2 + a_{23}x_2x_3$$
$$+ a_{31}x_3x_1 + a_{32}x_3x_2 + a_{33}x_3x_3$$
$$= a_{11}x_1^2 + a_{22}x_2^2 + a_{33}x_3^2$$
$$+ (a_{13} + a_{31})x_1x_3 + (a_{12} + a_{21})x_1x_2 + (a_{23} + a_{32})x_3x_2.$$

26.2 Write in suffix notation div $\mathbf{A}$, $|\mathbf{A}|$, grad ϕ, div grad ϕ, where $\mathbf{A}$ is an arbitrary vector and ϕ is a scalar function (Cartesian axes x_1, x_2, x_3 assumed).

$$\text{div } \mathbf{A} = \frac{\partial A_1}{\partial x_1} + \frac{\partial A_2}{\partial x_2} + \frac{\partial A_3}{\partial x_3} = \frac{\partial A_k}{\partial x_k} \quad \text{(summation over } k = 1, 2, 3)$$

$$|\mathbf{A}| = \sqrt{(A_1^2 + A_2^2 + A_3^2)} = \sqrt{A_kA_k} \quad \text{(summation over } k = 1, 2, 3).$$

$$\text{grad } \phi = \mathbf{i}\frac{\partial \phi}{\partial x_1} + \mathbf{j}\frac{\partial \phi}{\partial x_2} + \mathbf{k}\frac{\partial \phi}{\partial x_3}; \quad i\text{th component: } (\text{grad } \phi)_i = \partial\phi/\partial x_i.$$

$$\text{div grad } \phi = \frac{\partial}{\partial x_1}\left(\frac{\partial \phi}{\partial x_1}\right) + \frac{\partial}{\partial x_2}\left(\frac{\partial \phi}{\partial x_2}\right)$$

$$+ \frac{\partial}{\partial x_3}\left(\frac{\partial \phi}{\partial x_3}\right) = \frac{\partial^2 \phi}{\partial x_i \partial x_i}$$

26.3 Find the values of $\delta_{ij}\delta_{ij}$, $\delta_{ij}\delta_{jk}\delta_{km}\delta_{im}$, $\varepsilon_{ijk}A_jA_k$, $\delta_{ij}\partial A_i/\partial x_j$, $\delta_{ik}\varepsilon_{ikm}$ (all indices take values $1, 2, 3$).

$\delta_{ij} = 0$, $i \neq j$; $\delta_{ij} = 1$, $i = j$.

$\varepsilon_{ijk} = 0$ if any pair of indices are the same. $\varepsilon_{ijk} = -\varepsilon_{jik}$, etc.

$\varepsilon_{123} = 1$, $\varepsilon_{ijk} = \varepsilon_{jki} = \varepsilon_{kij}$.

$\delta_{ij}\delta_{ij} = \delta_{11}\delta_{11} + \delta_{22}\delta_{22} + \delta_{33}\delta_{33} = 1 + 1 + 1 = 3$.

$\delta_{ij}\delta_{jk}\delta_{km}\delta_{im} = \delta_{ik}\delta_{ki} = \delta_{ii} = 3$.

$\varepsilon_{ijk}A_jA_k$ (summation over j, k). Terms with $j = k$ are zero since $\varepsilon_{ijk} = 0$ if any pair of indices are the same. Now take $i = 3$, say. Then

$$\varepsilon_{3jk}A_jA_k = \varepsilon_{312}A_1A_2 + \varepsilon_{321}A_2A_1 = \varepsilon_{312}(A_1A_2 - A_2A_1) = 0.$$

So also for $i = 1, 2$. Hence $\varepsilon_{ijk}A_jA_k = 0$.

Alternatively we notice that ε_{ijk} is skew-symmetric in any pair of indices, whereas $A_j A_k$ is symmetric. It follows directly that $\varepsilon_{ijk} A_j A_k = 0$.

$$\delta_{ij} \frac{\partial A_i}{\partial x_j} = \frac{\partial A_i}{\partial x_i} = \text{div } \mathbf{A}.$$

$$\delta_{ik} \varepsilon_{ikm} = \varepsilon_{iim} = 0.$$

26.4 Evaluate $\varepsilon_{ikl} \varepsilon_{jkl}, \ \varepsilon_{jkl} \varepsilon_{klj}$.

Basic result

$\varepsilon_{ikl} \varepsilon_{jml} = \delta_{ij} \delta_{km} - \delta_{jk} \delta_{im}$ (summation over repeated indices as usual). Hence

$$\varepsilon_{ikl} \varepsilon_{jkl} = \delta_{ij} \delta_{kk} - \delta_{jk} \delta_{ik} = 3\delta_{ij} - \delta_{ij} = 2\delta_{ij},$$

(using $\delta_{kk} = \delta_{11} + \delta_{22} + \delta_{33} = 3$). Also

$$\varepsilon_{jkl} \varepsilon_{klj} = \varepsilon_{jkl} \varepsilon_{jkl} = 2\delta_{ii} = 6.$$

26.5 Use the permutation symbol ε_{ijk} to show that

curl curl $\mathbf{A} = \text{grad div } \mathbf{A} - \nabla^2 \mathbf{A}$.

Basic result

From the definition of curl $\mathbf{A}$ (see 20.6), we can write the ith component of curl $\mathbf{A}$ in terms of the permutation symbol as

$$(\text{curl } \mathbf{A})_i = \varepsilon_{ijk} \frac{\partial A_k}{\partial x_j}$$

(remembering that summation occurs over repeated indices). For example, the 1st component of curl $\mathbf{A}$ is

$$\varepsilon_{1jk} \frac{\partial A_k}{\partial x_j} = \varepsilon_{123} \frac{\partial A_3}{\partial x_2} + \varepsilon_{132} \frac{\partial A_2}{\partial x_3}$$

$$= \varepsilon_{123} \left(\frac{\partial A_3}{\partial x_2} - \frac{\partial A_2}{\partial x_3} \right) = \left(\frac{\partial A_3}{\partial x_2} - \frac{\partial A_2}{\partial x_3} \right)$$

Now consider the pth component ($p = 1, 2$ or 3)

$$(\text{curl curl } \mathbf{A})_p = \varepsilon_{pqr} \frac{\partial}{\partial x_q} \left(\varepsilon_{rjk} \frac{\partial A_k}{\partial x_j} \right)$$

$$= \varepsilon_{rpq} \varepsilon_{rjk} \frac{\partial^2 A_k}{\partial x_q \, \partial x_j}$$

$$= (\delta_{pj} \delta_{qk} - \delta_{pk} \delta_{qj}) \frac{\partial^2 A_k}{\partial x_q \, \partial x_j}$$

$$= \frac{\partial^2 A_k}{\partial x_k \, \partial x_p} - \frac{\partial^2 A_p}{\partial x_j \, \partial x_j}.$$

$$(\text{grad div } \mathbf{A})_p = \frac{\partial}{\partial x_p} \left(\frac{\partial A_k}{\partial x_k} \right) = \frac{\partial^2 A_k}{\partial x_k \, \partial x_p}. \qquad (\nabla^2 \mathbf{A})_p = \frac{\partial^2 A_p}{\partial x_k \, \partial x_k}.$$

Hence

$$(\text{curl curl } \mathbf{A})_p = (\text{grad div } \mathbf{A})_p - (\nabla^2 \mathbf{A})_p.$$

Since this result is true for $p = 1, 2, 3$ it is true as a vector result.

26.6 Given $\phi = a_{ik} x_i x_k$, obtain an expression for $\partial\phi/\partial x_s$, where a_{ik} are constants.

$$\frac{\partial \phi}{\partial x_s} = a_{ik} \left(\frac{\partial x_i}{\partial x_s} x_k + x_i \frac{\partial x_k}{\partial x_s} \right) = a_{ik} \delta_{is} x_k + a_{ik} x_i \delta_{ks}$$

$$= a_{sk} x_k + a_{is} x_i = (a_{si} + a_{is}) x_i.$$

26.7 Show that the ith component of any vector of the form $\mathbf{a} \wedge (\mathbf{b} \wedge \mathbf{a})$ may be written in the form $c_{ij} b_j$, where c_{ij} is symmetric. Find the elements of c_{ij}.

Now the jth component of the vector product $\mathbf{b} \wedge \mathbf{a}$ is $(\mathbf{b} \wedge \mathbf{a})_j = \varepsilon_{jkl} b_k a_l$. Therefore

$$[\mathbf{a} \wedge (\mathbf{b} \wedge \mathbf{a})]_i = \varepsilon_{imn} a_m \varepsilon_{nkl} b_k a_l$$

$$= \varepsilon_{nim} \varepsilon_{nkl} a_m b_k a_l$$

$$= (\delta_{ik} \delta_{ml} - \delta_{il} \delta_{mk}) a_m b_k a_l$$

$$= (a_m a_m \delta_{ik} - a_i a_k) b_k$$

$$= c_{ik} b_k, \quad \text{where} \quad c_{ik} = a_m a_m \delta_{ik} - a_i a_k.$$

Clearly $c_{ki} = c_{ik}$, so c_{ik} is symmetric.

26.8 Prove that if $\phi(x_i)$ satisfies the equation $\nabla^2 \phi + \omega^2 \phi = 0$, where ω is a constant, then the second-order tensor

$$G_{ln} = (\nabla_l \nabla_n - \delta_{ln} \nabla^2) \phi$$

is a solution of the equation

$$\varepsilon_{ijk}\varepsilon_{lmi} \nabla_j \nabla_m G_{kn} - \omega^2 G_{ln} = 0.$$

Consider

$$\varepsilon_{ijk}\varepsilon_{lmi} \nabla_j \nabla_m G_{kn} = \varepsilon_{ijk}\varepsilon_{lmi} \nabla_j \nabla_m [\nabla_k \nabla_n - \delta_{kn} \nabla^2]\phi.$$

The first term on the right is

$$\varepsilon_{ijk}\varepsilon_{lmi} \nabla_j \nabla_m \nabla_k \nabla_n\phi$$
$$= \varepsilon_{jki}\varepsilon_{lmi} \nabla_j \nabla_m \nabla_k \nabla_n\phi$$
$$= (\delta_{jl} \delta_{km} - \delta_{jm} \delta_{kl})\nabla_j \nabla_m \nabla_k \nabla_n\phi$$
$$= \nabla_l \nabla_k \nabla_k \nabla_n\phi - \nabla_m \nabla_m \nabla_l \nabla_n\phi$$
$$= \nabla_l \nabla^2 \nabla_n\phi - \nabla^2 \nabla_l \nabla_n\phi$$
$$= \nabla^2(\nabla_l \nabla_n - \nabla_l \nabla_n)\phi = 0.$$

The second term on the right is

$$- \varepsilon_{ijk}\varepsilon_{lmi} \delta_{kn} \nabla_j \nabla_m \nabla^2\phi$$
$$= -\varepsilon_{ijn}\varepsilon_{lmi} \nabla_j \nabla_m \nabla^2\phi = -\varepsilon_{jni}\varepsilon_{lmi} \nabla_j \nabla_m \nabla^2\phi$$
$$= -(\delta_{jl} \delta_{nm} - \delta_{jm} \delta_{nl}) \nabla_j \nabla_m \nabla^2\phi = -\nabla_l \nabla_n \nabla^2\phi$$
$$+ \delta_{nl} \nabla^2(\nabla^2\phi).$$

For the equation to be satisfied therefore this should be put equal to $\omega^2 G_{ln}$. But $\omega^2 G_{ln} = \omega^2(\nabla_l \nabla_n\phi - \delta_{ln} \nabla^2\phi)$ (from the definition of G_{ln}). Hence we have

$$-(\nabla_l \nabla_n \nabla^2\phi - \delta_{ln} \nabla^2(\nabla^2\phi)) = \omega^2 \nabla_l \nabla_n\phi - \omega^2 \delta_{ln} \nabla^2\phi$$

or

$$\nabla_l \nabla_n(\nabla^2\phi + \omega^2\phi) - \delta_{ln} \nabla^2(\nabla^2\phi + \omega^2\phi) = 0$$

which is satisfied if $\nabla^2\phi + \omega^2\phi = 0$.

26.9 Show that if $A_k = x_i x_i \varepsilon_{ijk} + x_i x_i x_k$, then

$$\frac{\partial A_k}{\partial x^s} = 2x_s x_k + x_i x_i \delta_{ks}$$

find also $\partial^2 A_k/\partial x_r \partial x_s$, and show that $\partial^2 A_k/\partial x_s^2 = 2x_k + 4 \delta_{ks}x_s$, where there is *no* summation over the index s.

$$\frac{\partial A_k}{\partial x^s} = \frac{\partial x_i}{\partial x_s} x_j \varepsilon_{ijk} + x_i \frac{\partial x_j}{\partial x_s} \varepsilon_{ijk}$$

$$+ x_i x_j \frac{\partial}{\partial x_s} \varepsilon_{ijk} + 2 \frac{\partial x_i}{\partial x_s} x_i x_k + x_i x_i \frac{\partial x_k}{\partial x_s}$$

$$= \delta_{is} x_j \varepsilon_{ijk} + x_i \delta_{js} \varepsilon_{ijk} + 0 + 2 x_i x_k \delta_{is} + x_i x_i \delta_{ks}$$

$$= x_j \varepsilon_{sjk} + x_i \varepsilon_{isk} + 2 x_s x_k + x_i x_i \delta_{ks}$$

$$= x_i (\varepsilon_{sik} + \varepsilon_{isk}) + 2 x_s x_k + x_i x_i \delta_{ks}$$

$$= 0 + 2 x_s x_k + x_i x_i \delta_{ks}.$$

$$\frac{\partial^2 A_k}{\partial x_r \, \partial x_s} = \frac{\partial}{\partial x_r} \left(\frac{\partial A_k}{\partial x_s} \right) = \frac{\partial}{\partial x_r} (2 x_s x_k + x_i x_i \delta_{ks})$$

$$= 2 \left(\frac{\partial x_s}{\partial x_r} x_k + x_s \frac{\partial x_k}{\partial x_r} \right) + 2 x_i \frac{\partial x_i}{\partial x_r} \delta_{ks}$$

$$= 2 \, \delta_{rs} x_k + 2 x_s \, \delta_{kr} + 2 x_i \, \delta_{ir} \, \delta_{ks}$$

$$= 2 (x_k \, \delta_{rs} + x_s \, \delta_{kr} + x_r \, \delta_{ks}).$$

If $r = s$ (no summation over s) then

$$\frac{\partial^2 A_k}{\partial x_s^2} = 2 (x_k \, \delta_{ss} + x_s \, \delta_{ks} + x_s \, \delta_{ks})$$

$$= 2 x_k + 4 x_s \, \delta_{ks} \qquad \text{(since } \delta_{ss} = 1 \text{)}.$$

(Problem contributed by Dr. Tony Dowson)

A CATALOG OF SELECTED
DOVER BOOKS
IN SCIENCE AND MATHEMATICS

Engineering

FUNDAMENTALS OF ASTRODYNAMICS, Roger R. Bate, Donald D. Mueller, and Jerry E. White. Teaching text developed by U.S. Air Force Academy develops the basic two-body and n-body equations of motion; orbit determination; classical orbital elements, coordinate transformations; differential correction; more. 1971 edition. 455pp. 5 3/8 x 8 1/2. 0-486-60061-0

INTRODUCTION TO CONTINUUM MECHANICS FOR ENGINEERS: Revised Edition, Ray M. Bowen. This self-contained text introduces classical continuum models within a modern framework. Its numerous exercises illustrate the governing principles, linearizations, and other approximations that constitute classical continuum models. 2007 edition. 320pp. 6 1/8 x 9 1/4. 0-486-47460-7

ENGINEERING MECHANICS FOR STRUCTURES, Louis L. Bucciarelli. This text explores the mechanics of solids and statics as well as the strength of materials and elasticity theory. Its many design exercises encourage creative initiative and systems thinking. 2009 edition. 320pp. 6 1/8 x 9 1/4. 0-486-46855-0

FEEDBACK CONTROL THEORY, John C. Doyle, Bruce A. Francis and Allen R. Tannenbaum. This excellent introduction to feedback control system design offers a theoretical approach that captures the essential issues and can be applied to a wide range of practical problems. 1992 edition. 224pp. 6 1/2 x 9 1/4. 0-486-46933-6

THE FORCES OF MATTER, Michael Faraday. These lectures by a famous inventor offer an easy-to-understand introduction to the interactions of the universe's physical forces. Six essays explore gravitation, cohesion, chemical affinity, heat, magnetism, and electricity. 1993 edition. 96pp. 5 3/8 x 8 1/2. 0-486-47482-8

DYNAMICS, Lawrence E. Goodman and William H. Warner. Beginning engineering text introduces calculus of vectors, particle motion, dynamics of particle systems and plane rigid bodies, technical applications in plane motions, and more. Exercises and answers in every chapter. 619pp. 5 3/8 x 8 1/2. 0-486-42006-X

ADAPTIVE FILTERING PREDICTION AND CONTROL, Graham C. Goodwin and Kwai Sang Sin. This unified survey focuses on linear discrete-time systems and explores natural extensions to nonlinear systems. It emphasizes discrete-time systems, summarizing theoretical and practical aspects of a large class of adaptive algorithms. 1984 edition. 560pp. 6 1/2 x 9 1/4. 0-486-46932-8

INDUCTANCE CALCULATIONS, Frederick W. Grover. This authoritative reference enables the design of virtually every type of inductor. It features a single simple formula for each type of inductor, together with tables containing essential numerical factors. 1946 edition. 304pp. 5 3/8 x 8 1/2. 0-486-47440-2

THERMODYNAMICS: Foundations and Applications, Elias P. Gyftopoulos and Gian Paolo Beretta. Designed by two MIT professors, this authoritative text discusses basic concepts and applications in detail, emphasizing generality, definitions, and logical consistency. More than 300 solved problems cover realistic energy systems and processes. 800pp. 6 1/8 x 9 1/4. 0-486-43932-1

THE FINITE ELEMENT METHOD: Linear Static and Dynamic Finite Element Analysis, Thomas J. R. Hughes. Text for students without in-depth mathematical training, this text includes a comprehensive presentation and analysis of algorithms of time-dependent phenomena plus beam, plate, and shell theories. Solution guide available upon request. 672pp. 6 1/2 x 9 1/4. 0-486-41181-8

HELICOPTER THEORY, Wayne Johnson. Monumental engineering text covers vertical flight, forward flight, performance, mathematics of rotating systems, rotary wing dynamics and aerodynamics, aeroelasticity, stability and control, stall, noise, and more. 189 illustrations. 1980 edition. 1089pp. 5 5/8 x 8 1/4.   0-486-68230-7

MATHEMATICAL HANDBOOK FOR SCIENTISTS AND ENGINEERS: Definitions, Theorems, and Formulas for Reference and Review, Granino A. Korn and Theresa M. Korn. Convenient access to information from every area of mathematics: Fourier transforms, Z transforms, linear and nonlinear programming, calculus of variations, random-process theory, special functions, combinatorial analysis, game theory, much more. 1152pp. 5 3/8 x 8 1/2.   0-486-41147-8

A HEAT TRANSFER TEXTBOOK: Fourth Edition, John H. Lienhard V and John H. Lienhard IV. This introduction to heat and mass transfer for engineering students features worked examples and end-of-chapter exercises. Worked examples and end-of-chapter exercises appear throughout the book, along with well-drawn, illuminating figures. 768pp. 7 x 9 1/4.   0-486-47931-5

BASIC ELECTRICITY, U.S. Bureau of Naval Personnel. Originally a training course; best nontechnical coverage. Topics include batteries, circuits, conductors, AC and DC, inductance and capacitance, generators, motors, transformers, amplifiers, etc. Many questions with answers. 349 illustrations. 1969 edition. 448pp. 6 1/2 x 9 1/4.
0-486-20973-3

BASIC ELECTRONICS, U.S. Bureau of Naval Personnel. Clear, well-illustrated introduction to electronic equipment covers numerous essential topics: electron tubes, semiconductors, electronic power supplies, tuned circuits, amplifiers, receivers, ranging and navigation systems, computers, antennas, more. 560 illustrations. 567pp. 6 1/2 x 9 1/4.   0-486-21076-6

BASIC WING AND AIRFOIL THEORY, Alan Pope. This self-contained treatment by a pioneer in the study of wind effects covers flow functions, airfoil construction and pressure distribution, finite and monoplane wings, and many other subjects. 1951 edition. 320pp. 5 3/8 x 8 1/2.   0-486-47188-8

SYNTHETIC FUELS, Ronald F. Probstein and R. Edwin Hicks. This unified presentation examines the methods and processes for converting coal, oil, shale, tar sands, and various forms of biomass into liquid, gaseous, and clean solid fuels. 1982 edition. 512pp. 6 1/8 x 9 1/4.   0-486-44977-7

THEORY OF ELASTIC STABILITY, Stephen P. Timoshenko and James M. Gere. Written by world-renowned authorities on mechanics, this classic ranges from theoretical explanations of 2- and 3-D stress and strain to practical applications such as torsion, bending, and thermal stress. 1961 edition. 560pp. 5 3/8 x 8 1/2.   0-486-47207-8

PRINCIPLES OF DIGITAL COMMUNICATION AND CODING, Andrew J. Viterbi and Jim K. Omura. This classic by two digital communications experts is geared toward students of communications theory and to designers of channels, links, terminals, modems, or networks used to transmit and receive digital messages. 1979 edition. 576pp. 6 1/8 x 9 1/4.   0-486-46901-8

LINEAR SYSTEM THEORY: The State Space Approach, Lotfi A. Zadeh and Charles A. Desoer. Written by two pioneers in the field, this exploration of the state space approach focuses on problems of stability and control, plus connections between this approach and classical techniques. 1963 edition. 656pp. 6 1/8 x 9 1/4.
0-486-46663-9

Browse over 9,000 books at www.doverpublications.com

Physics

THEORETICAL NUCLEAR PHYSICS, John M. Blatt and Victor F. Weisskopf. An uncommonly clear and cogent investigation and correlation of key aspects of theoretical nuclear physics by leading experts: the nucleus, nuclear forces, nuclear spectroscopy, two-, three- and four-body problems, nuclear reactions, beta-decay and nuclear shell structure. 896pp. 5 3/8 x 8 1/2. 0-486-66827-4

QUANTUM THEORY, David Bohm. This advanced undergraduate-level text presents the quantum theory in terms of qualitative and imaginative concepts, followed by specific applications worked out in mathematical detail. 655pp. 5 3/8 x 8 1/2. 0-486-65969-0

ATOMIC PHYSICS AND HUMAN KNOWLEDGE, Niels Bohr. Articles and speeches by the Nobel Prize–winning physicist, dating from 1934 to 1958, offer philosophical explorations of the relevance of atomic physics to many areas of human endeavor. 1961 edition. 112pp. 5 3/8 x 8 1/2. 0-486-47928-5

COSMOLOGY, Hermann Bondi. A co-developer of the steady-state theory explores his conception of the expanding universe. This historic book was among the first to present cosmology as a separate branch of physics. 1961 edition. 192pp. 5 3/8 x 8 1/2. 0-486-47483-6

LECTURES ON QUANTUM MECHANICS, Paul A. M. Dirac. Four concise, brilliant lectures on mathematical methods in quantum mechanics from Nobel Prize–winning quantum pioneer build on idea of visualizing quantum theory through the use of classical mechanics. 96pp. 5 3/8 x 8 1/2. 0-486-41713-1

THE PRINCIPLE OF RELATIVITY, Albert Einstein and Frances A. Davis. Eleven papers that forged the general and special theories of relativity include seven papers by Einstein, two by Lorentz, and one each by Minkowski and Weyl. 1923 edition. 240pp. 5 3/8 x 8 1/2. 0-486-60081-5

PHYSICS OF WAVES, William C. Elmore and Mark A. Heald. Ideal as a classroom text or for individual study, this unique one-volume overview of classical wave theory covers wave phenomena of acoustics, optics, electromagnetic radiations, and more. 477pp. 5 3/8 x 8 1/2. 0-486-64926-1

THERMODYNAMICS, Enrico Fermi. In this classic of modern science, the Nobel Laureate presents a clear treatment of systems, the First and Second Laws of Thermodynamics, entropy, thermodynamic potentials, and much more. Calculus required. 160pp. 5 3/8 x 8 1/2. 0-486-60361-X

QUANTUM THEORY OF MANY-PARTICLE SYSTEMS, Alexander L. Fetter and John Dirk Walecka. Self-contained treatment of nonrelativistic many-particle systems discusses both formalism and applications in terms of ground-state (zero-temperature) formalism, finite-temperature formalism, canonical transformations, and applications to physical systems. 1971 edition. 640pp. 5 3/8 x 8 1/2. 0-486-42827-3

QUANTUM MECHANICS AND PATH INTEGRALS: Emended Edition, Richard P. Feynman and Albert R. Hibbs. Emended by Daniel F. Styer. The Nobel Prize–winning physicist presents unique insights into his theory and its applications. Feynman starts with fundamentals and advances to the perturbation method, quantum electrodynamics, and statistical mechanics. 1965 edition, emended in 2005. 384pp. 6 1/8 x 9 1/4. 0-486-47722-3

Browse over 9,000 books at www.doverpublications.com

Physics

INTRODUCTION TO MODERN OPTICS, Grant R. Fowles. A complete basic undergraduate course in modern optics for students in physics, technology, and engineering. The first half deals with classical physical optics; the second, quantum nature of light. Solutions. 336pp. 5 3/8 x 8 1/2. 0-486-65957-7

THE QUANTUM THEORY OF RADIATION: Third Edition, W. Heitler. The first comprehensive treatment of quantum physics in any language, this classic introduction to basic theory remains highly recommended and widely used, both as a text and as a reference. 1954 edition. 464pp. 5 3/8 x 8 1/2. 0-486-64558-4

QUANTUM FIELD THEORY, Claude Itzykson and Jean-Bernard Zuber. This comprehensive text begins with the standard quantization of electrodynamics and perturbative renormalization, advancing to functional methods, relativistic bound states, broken symmetries, nonabelian gauge fields, and asymptotic behavior. 1980 edition. 752pp. 6 1/2 x 9 1/4. 0-486-44568-2

FOUNDATIONS OF POTENTIAL THERY, Oliver D. Kellogg. Introduction to fundamentals of potential functions covers the force of gravity, fields of force, potentials, harmonic functions, electric images and Green's function, sequences of harmonic functions, fundamental existence theorems, and much more. 400pp. 5 3/8 x 8 1/2. 0-486-60144-7

FUNDAMENTALS OF MATHEMATICAL PHYSICS, Edgar A. Kraut. Indispensable for students of modern physics, this text provides the necessary background in mathematics to study the concepts of electromagnetic theory and quantum mechanics. 1967 edition. 480pp. 6 1/2 x 9 1/4. 0-486-45809-1

GEOMETRY AND LIGHT: The Science of Invisibility, Ulf Leonhardt and Thomas Philbin. Suitable for advanced undergraduate and graduate students of engineering, physics, and mathematics and scientific researchers of all types, this is the first authoritative text on invisibility and the science behind it. More than 100 full-color illustrations, plus exercises with solutions. 2010 edition. 288pp. 7 x 9 1/4. 0-486-47693-6

QUANTUM MECHANICS: New Approaches to Selected Topics, Harry J. Lipkin. Acclaimed as "excellent" (Nature) and "very original and refreshing" (Physics Today), these studies examine the Mössbauer effect, many-body quantum mechanics, scattering theory, Feynman diagrams, and relativistic quantum mechanics. 1973 edition. 480pp. 5 3/8 x 8 1/2. 0-486-45893-8

THEORY OF HEAT, James Clerk Maxwell. This classic sets forth the fundamentals of thermodynamics and kinetic theory simply enough to be understood by beginners, yet with enough subtlety to appeal to more advanced readers, too. 352pp. 5 3/8 x 8 1/2. 0-486-41735-2

QUANTUM MECHANICS, Albert Messiah. Subjects include formalism and its interpretation, analysis of simple systems, symmetries and invariance, methods of approximation, elements of relativistic quantum mechanics, much more. "Strongly recommended." – American Journal of Physics. 1152pp. 5 3/8 x 8 1/2. 0-486-40924-4

RELATIVISTIC QUANTUM FIELDS, Charles Nash. This graduate-level text contains techniques for performing calculations in quantum field theory. It focuses chiefly on the dimensional method and the renormalization group methods. Additional topics include functional integration and differentiation. 1978 edition. 240pp. 5 3/8 x 8 1/2. 0-486-47752-5

Browse over 9,000 books at www.doverpublications.com

Physics

MATHEMATICAL TOOLS FOR PHYSICS, James Nearing. Encouraging students' development of intuition, this original work begins with a review of basic mathematics and advances to infinite series, complex algebra, differential equations, Fourier series, and more. 2010 edition. 496pp. 6 1/8 x 9 1/4. 0-486-48212-X

TREATISE ON THERMODYNAMICS, Max Planck. Great classic, still one of the best introductions to thermodynamics. Fundamentals, first and second principles of thermodynamics, applications to special states of equilibrium, more. Numerous worked examples. 1917 edition. 297pp. 5 3/8 x 8. 0-486-66371-X

AN INTRODUCTION TO RELATIVISTIC QUANTUM FIELD THEORY, Silvan S. Schweber. Complete, systematic, and self-contained, this text introduces modern quantum field theory. "Combines thorough knowledge with a high degree of didactic ability and a delightful style." – Mathematical Reviews. 1961 edition. 928pp. 5 3/8 x 8 1/2. 0-486-44228-4

THE ELECTROMAGNETIC FIELD, Albert Shadowitz. Comprehensive undergraduate text covers basics of electric and magnetic fields, building up to electromagnetic theory. Related topics include relativity theory. Over 900 problems, some with solutions. 1975 edition. 768pp. 5 5/8 x 8 1/4. 0-486-65660-8

THE PRINCIPLES OF STATISTICAL MECHANICS, Richard C. Tolman. Definitive treatise offers a concise exposition of classical statistical mechanics and a thorough elucidation of quantum statistical mechanics, plus applications of statistical mechanics to thermodynamic behavior. 1930 edition. 704pp. 5 5/8 x 8 1/4. 0-486-63896-0

INTRODUCTION TO THE PHYSICS OF FLUIDS AND SOLIDS, James S. Trefil. This interesting, informative survey by a well-known science author ranges from classical physics and geophysical topics, from the rings of Saturn and the rotation of the galaxy to underground nuclear tests. 1975 edition. 320pp. 5 3/8 x 8 1/2. 0-486-47437-2

STATISTICAL PHYSICS, Gregory H. Wannier. Classic text combines thermodynamics, statistical mechanics, and kinetic theory in one unified presentation. Topics include equilibrium statistics of special systems, kinetic theory, transport coefficients, and fluctuations. Problems with solutions. 1966 edition. 532pp. 5 3/8 x 8 1/2. 0-486-65401-X

SPACE, TIME, MATTER, Hermann Weyl. Excellent introduction probes deeply into Euclidean space, Riemann's space, Einstein's general relativity, gravitational waves and energy, and laws of conservation. "A classic of physics." – British Journal for Philosophy and Science. 330pp. 5 3/8 x 8 1/2. 0-486-60267-2

RANDOM VIBRATIONS: Theory and Practice, Paul H. Wirsching, Thomas L. Paez and Keith Ortiz. Comprehensive text and reference covers topics in probability, statistics, and random processes, plus methods for analyzing and controlling random vibrations. Suitable for graduate students and mechanical, structural, and aerospace engineers. 1995 edition. 464pp. 5 3/8 x 8 1/2. 0-486-45015-5

PHYSICS OF SHOCK WAVES AND HIGH-TEMPERATURE HYDRO DYNAMIC PHENOMENA, Ya B. Zel'dovich and Yu P. Raizer. Physical, chemical processes in gases at high temperatures are focus of outstanding text, which combines material from gas dynamics, shock-wave theory, thermodynamics and statistical physics, other fields. 284 illustrations. 1966–1967 edition. 944pp. 6 1/8 x 9 1/4. 0-486-42002-7

Browse over 9,000 books at www.doverpublications.com

Chemistry

MOLECULAR COLLISION THEORY, M. S. Child. This high-level monograph offers an analytical treatment of classical scattering by a central force, quantum scattering by a central force, elastic scattering phase shifts, and semi-classical elastic scattering. 1974 edition. 310pp. 5 3/8 x 8 1/2.   0-486-69437-2

HANDBOOK OF COMPUTATIONAL QUANTUM CHEMISTRY, David B. Cook. This comprehensive text provides upper-level undergraduates and graduate students with an accessible introduction to the implementation of quantum ideas in molecular modeling, exploring practical applications alongside theoretical explanations. 1998 edition. 832pp. 5 3/8 x 8 1/2.   0-486-44307-8

RADIOACTIVE SUBSTANCES, Marie Curie. The celebrated scientist's thesis, which directly preceded her 1903 Nobel Prize, discusses establishing atomic character of radioactivity; extraction from pitchblende of polonium and radium; isolation of pure radium chloride; more. 96pp. 5 3/8 x 8 1/2.   0-486-42550-9

CHEMICAL MAGIC, Leonard A. Ford. Classic guide provides intriguing entertainment while elucidating sound scientific principles, with more than 100 unusual stunts: cold fire, dust explosions, a nylon rope trick, a disappearing beaker, much more. 128pp. 5 3/8 x 8 1/2.   0-486-67628-5

ALCHEMY, E. J. Holmyard. Classic study by noted authority covers 2,000 years of alchemical history: religious, mystical overtones; apparatus; signs, symbols, and secret terms; advent of scientific method, much more. Illustrated. 320pp. 5 3/8 x 8 1/2.
0-486-26298-7

CHEMICAL KINETICS AND REACTION DYNAMICS, Paul L. Houston. This text teaches the principles underlying modern chemical kinetics in a clear, direct fashion, using several examples to enhance basic understanding. Solutions to selected problems. 2001 edition. 352pp. 8 3/8 x 11.   0-486-45334-0

PROBLEMS AND SOLUTIONS IN QUANTUM CHEMISTRY AND PHYSICS, Charles S. Johnson and Lee G. Pedersen. Unusually varied problems, with detailed solutions, cover of quantum mechanics, wave mechanics, angular momentum, molecular spectroscopy, scattering theory, more. 280 problems, plus 139 supplementary exercises. 430pp. 6 1/2 x 9 1/4.   0-486-65236-X

ELEMENTS OF CHEMISTRY, Antoine Lavoisier. Monumental classic by the founder of modern chemistry features first explicit statement of law of conservation of matter in chemical change, and more. Facsimile reprint of original (1790) Kerr translation. 539pp. 5 3/8 x 8 1/2.   0-486-64624-6

MAGNETISM AND TRANSITION METAL COMPLEXES, F. E. Mabbs and D. J. Machin. A detailed view of the calculation methods involved in the magnetic properties of transition metal complexes, this volume offers sufficient background for original work in the field. 1973 edition. 240pp. 5 3/8 x 8 1/2.   0-486-46284-6

GENERAL CHEMISTRY, Linus Pauling. Revised third edition of classic first-year text by Nobel laureate. Atomic and molecular structure, quantum mechanics, statistical mechanics, thermodynamics correlated with descriptive chemistry. Problems. 992pp. 5 3/8 x 8 1/2.   0-486-65622-5

ELECTROLYTE SOLUTIONS: Second Revised Edition, R. A. Robinson and R. H. Stokes. Classic text deals primarily with measurement, interpretation of conductance, chemical potential, and diffusion in electrolyte solutions. Detailed theoretical interpretations, plus extensive tables of thermodynamic and transport properties. 1970 edition. 590pp. 5 3/8 x 8 1/2.   0-486-42225-9

Browse over 9,000 books at www.doverpublications.com

CATALOG OF DOVER BOOKS

Mathematics–Logic and Problem Solving

PERPLEXING PUZZLES AND TANTALIZING TEASERS, Martin Gardner. Ninety-three riddles, mazes, illusions, tricky questions, word and picture puzzles, and other challenges offer hours of entertainment for youngsters. Filled with rib-tickling drawings. Solutions. 224pp. 5 3/8 x 8 1/2.   0-486-25637-5

MY BEST MATHEMATICAL AND LOGIC PUZZLES, Martin Gardner. The noted expert selects 70 of his favorite "short" puzzles. Includes The Returning Explorer, The Mutilated Chessboard, Scrambled Box Tops, and dozens more. Complete solutions included. 96pp. 5 3/8 x 8 1/2.   0-486-28152-3

THE LADY OR THE TIGER?: and Other Logic Puzzles, Raymond M. Smullyan. Created by a renowned puzzle master, these whimsically themed challenges involve paradoxes about probability, time, and change; metapuzzles; and self-referentiality. Nineteen chapters advance in difficulty from relatively simple to highly complex. 1982 edition. 240pp. 5 3/8 x 8 1/2.   0-486-47027-X

SATAN, CANTOR AND INFINITY: Mind-Boggling Puzzles, Raymond M. Smullyan. A renowned mathematician tells stories of knights and knaves in an entertaining look at the logical precepts behind infinity, probability, time, and change. Requires a strong background in mathematics. Complete solutions. 288pp. 5 3/8 x 8 1/2.
0-486-47036-9

THE RED BOOK OF MATHEMATICAL PROBLEMS, Kenneth S. Williams and Kenneth Hardy. Handy compilation of 100 practice problems, hints and solutions indispensable for students preparing for the William Lowell Putnam and other mathematical competitions. Preface to the First Edition. Sources. 1988 edition. 192pp. 5 3/8 x 8 1/2.   0-486-69415-1

KING ARTHUR IN SEARCH OF HIS DOG AND OTHER CURIOUS PUZZLES, Raymond M. Smullyan. This fanciful, original collection for readers of all ages features arithmetic puzzles, logic problems related to crime detection, and logic and arithmetic puzzles involving King Arthur and his Dogs of the Round Table. 160pp. 5 3/8 x 8 1/2.
0-486-47435-6

UNDECIDABLE THEORIES: Studies in Logic and the Foundation of Mathematics, Alfred Tarski in collaboration with Andrzej Mostowski and Raphael M. Robinson. This well-known book by the famed logician consists of three treatises: "A General Method in Proofs of Undecidability," "Undecidability and Essential Undecidability in Mathematics," and "Undecidability of the Elementary Theory of Groups." 1953 edition. 112pp. 5 3/8 x 8 1/2.   0-486-47703-7

LOGIC FOR MATHEMATICIANS, J. Barkley Rosser. Examination of essential topics and theorems assumes no background in logic. "Undoubtedly a major addition to the literature of mathematical logic." – *Bulletin of the American Mathematical Society.* 1978 edition. 592pp. 6 1/8 x 9 1/4.   0-486-46898-4

INTRODUCTION TO PROOF IN ABSTRACT MATHEMATICS, Andrew Wohlgemuth. This undergraduate text teaches students what constitutes an acceptable proof, and it develops their ability to do proofs of routine problems as well as those requiring creative insights. 1990 edition. 384pp. 6 1/2 x 9 1/4.   0-486-47854-8

FIRST COURSE IN MATHEMATICAL LOGIC, Patrick Suppes and Shirley Hill. Rigorous introduction is simple enough in presentation and context for wide range of students. Symbolizing sentences; logical inference; truth and validity; truth tables; terms, predicates, universal quantifiers; universal specification and laws of identity; more. 288pp. 5 3/8 x 8 1/2.   0-486-42259-3

Browse over 9,000 books at www.doverpublications.com